LOW HEAD HYDROPOWER FOR LOCAL ENERGY SOLUTIONS

LOW HEAD HYDROPOWER FOR LOCAL ENERGY SOLUTIONS

DISSERTATION

Submitted in fulfillment of the requirements of
the Board for Doctorates of Delft University of Technology
and
of the Academic Board of the UNESCO-IHE
Institute for Water Education
for
the Degree of DOCTOR
to be defended in public on
Monday, 9 October 2017, at 10:00 hours
in Delft, the Netherlands

by

Arcot Ganesh Pradeep NARRAIN
Master of Science in Water Resources Engineering and Management
University of Stuttgart, Germany

born in Bangalore, India

This dissertation has been approved by the promotors:
Prof.dr.ir. A.E. Mynett
Prof.dr. N.G. Wright

Composition of the doctoral committee:

Chairman	Rector Magnificus Delft University of Technology
Vice-Chairman	Rector UNESCO-IHE
Prof.dr.ir. A.E. Mynett	UNESCO-IHE / Delft University of Technology, promotor
Prof.dr. N.G. Wright	De Montfort University, UK / UNESCO-IHE, promotor

Independent members:
Prof.dr.ir. W.S.J. Uijttewaal	Delft University of Technology
Prof.dr.ir. C. Zevenbergen	UNESCO-IHE / Delft University of Technology
Prof. dr. G. Pender	Heriot-Watt University, UK
Prof.dr.-ing. U. Gärtner	Esslingen University of Applied Sciences, Germany
Prof.dr.ir. H.H.G. Savenije	Delft University of Technology, reserve member

CRC Press/Balkema is an imprint of the Taylor & Francis Group, an informa business

Published by:
CRC Press/Balkema
Schipolweg 107C, 2316 XC, Leiden, The Netherlands
Pub.NL@taylorandfrancis.com
www.crcpress.com – www.taylorandfrancis.com
ISBN 978-0-8153-9612-3

"The power of water has changed more in this world than emperors or kings."

Leonardo da Vinci, (1452 – 1519)

Summary

Growing energy demand together with increasing environmental concerns has widened the interest in hydropower generation. In addition to conventional large-scale dam-operated hydropower technologies, several smaller scale hydropower machines are being explored for providing local energy solutions. Advantages of such machines are that they are cheaper to produce and are likely to have less environmental impact. Such technologies seem of particular interest for developing countries where technical solutions rely on the schooling of communities to operate machines with a new technology. The implementation of such technologies requires trained staff that is to be employed on site.

The role of small hydropower is becoming increasingly important on a global level. The potential in terms of sustainability is rapidly expanding as fossil resources and nuclear fuels are being depleted. Rapid industrial and economic growth in developing countries is leading to shortages of electricity, in particular peak demand. Economic growth is often given priority over environmental issues. In developing countries, particularly in rural areas, energy availability is not receiving the priority it deserves. Hence the possibility of creating power generation for local consumption based on sustainable technologies can improve living conditions as well as bring employment benefits to local communities.

Off-grid generation brings along many advantages and offers more flexibility for power generation. Changing legislation facilitates small hydropower generation by individuals and communities. The cost factor plays an important role in the construction of electro-mechanical equipment. Local availability of construction materials is of considerable importance, as machines require maintenance and timely repair in case of failure. For that purpose, many remote areas will require training of local individuals. The United Nations program "Sustainable Development Goals" (Goal #6: Water for all) and (Goal #7: Energy for Development) are of particular interest for the development of small-scale hydropower, as the effects of energy production in remote areas are of considerable importance for the local populace. The SDG's aim at alleviating poverty by providing basic amenities for all in the fields of health, education, clean water and sanitation, employment, affordable clean energy, etc.

In this research, the application of new hydropower technologies, in particular the local potential of water wheels in developing countries was studied. Increased interest in small hydropower as a renewable source of energy especially in regions of Asia and Africa, where the infrastructure is weak, have a potential for local power generation. The development of hydropower machines for very low heads was investigated in the EU-project HYLOW, "Hydropower converters for very low head differences" (HYLOW, 2012). The scope of the project was to develop two novel hydropower converters with free water surfaces and one in a closed system. These machines were to offer alternatives to conventional turbine technology in terms of environmental sustainability for very low head differences.

This thesis presents the approach of using numerical modelling of a small-scale hydropower machine. Such machines, like water wheels, have the advantage that the infrastructure around the deployment site does not require far-reaching incursions in the surroundings at the watercourse. The focus in this study is on improving the hydraulic efficiency. The torque thus generated can be converted into electrical or mechanical energy. The use of numerical models to assess and predict performance is a method used in many fields today. The increase in computational hardware and Computational Fluid Dynamics (CFD) simulation software nowadays enables numerical models to be more accurate and incorporate more physical complexities like free surfaces in rotating machines.

Various parameters such as choice of mesh elements, mesh consistency, choice of boundary conditions and turbulence model are studied in this thesis. To ensure that the boundary conditions were realistic, experimental results performed on a test rig within the HYLOW project were used to verify the numerical models. Flow measurements from the test rig were taken as real boundary conditions for the numerical simulations. The simulation results were checked and used for further developing the design of the hydropower machines. This approach enables obtaining a better understanding of flow conditions during operation by visualising flow patterns. The basic model supplied results that were plausible and matched those of the experiments, enabling modifications of the machine geometry e.g. blade angles to be made in the simulation models. These models could then be modified to further explore improving efficiency.

The numerical models were developed using the commercial CFD-code ANSYS Fluent®. The results of the simulations show that a change in blade geometry results in higher values for torque in the selected cases, thereby increasing the performance of the machine. The analysis of the flow situation shows that losses due to turbulence occur within the channel before and after the wheel, reducing the overall performance of the machine. To investigate the effect of channel-width, the model was modified by reducing the wheel-width within the given channel-width of 1 metre. The geometry of the new model called for a different meshing strategy consisting of tetrahedral cells with regions of local refinement. A complete re-meshing of the models with varying wheel-width was performed. The simulations show an increase in efficiency of the machine. Based on these simulations the optimal ratio of wheel-width to channel-width could be determined.

Modifications of the channel bed upstream and downstream of the machine show that the channel bed has an influence on the performance. Here too performance was improved. The influence of wall gaps at the wheel sides show that performance could be improved by modifying the gaps. Simulations show that variations in the downstream channel walls lead to changes in efficiency.

The results of the numerical simulations describe the flow conditions within the channel for different modifications in the channel and at the wheel. Power generation with these machines for low discharges is in the low kilowatt range, enabling the implementation of water wheels in areas with limited infrastructure. Hydropower machines like improved water wheels are seen to have potential for small-scale hydropower and seem to have value for local energy generation, in accordance with the Sustainable Development Goals (SDGs) of the United Nations.

Samenvatting

De steeds toenemende vraag naar energie en de groeiende aandacht voor het milieu hebben de belangstelling voor waterkrachtcentrales doen toenemen. Naast de ontwikkeling van grootschalige waterkrachtcentrales in bergachtige gebieden wordt er ook weer opnieuw gedacht over kleinschalige mogelijkheden op lokale schaal, waaronder de watermolen of het waterrad. De voordelen van dergelijke eenvoudige machines zijn dat ze goedkoper zijn te produceren en onderhouden, en minder effecten op het milieu hebben. Dit lijk met name van belang voor ontwikkelingslanden waar technische oplossingen kritisch afhankelijk zijn van de aanwezigheid van geschoolde arbeidskrachten die de nieuwe technologieën moeten toepassen.

Ook op wereldniveau neemt de belangstelling naar kleinschalige energiewinning uit waterkracht toe. Met de afname van fossiele en nucleaire brandstof neemt de belangstelling voor duurzame energiebronnen toe. De snelle toename van vraag naar energie in ontwikkelingslanden leidt tot een tekort aan beschikbare elektriciteit, met name tijdens piekuren. Economische groei krijgt veelal prioriteit boven milieuaspecten, met alle gevolgen van dien. In ontwikkelingslanden, met name in landelijke gebieden, krijgt energievoorziening niet de aandacht die het verdient. Juist daar kan de beschikbaarheid van duurzame energie het verschil maken tussen overleven of het opbouwen van een bestaan voor lokale gemeenschappen.

Het plaatselijk opwekken van energie biedt veel voordelen en flexibiliteit. Aanpassingen in wetgeving maken het mogelijk om kleinschalige waterkracht verder te ontwikkelen en testen. Uiteraard spelen kosten een belangrijke rol bij het ontwikkelen en bouwen van elektromechanische apparatuur. Plaatselijke beschikbaarheid van constructiematerialen is van groot belang voor het onderhouden en repareren van onderdelen. Daarbij is opleiding en training van de plaatselijke bevolking noodzakelijk. Het Verenigde Naties programma "Duurzame Ontwikkelingsdoelstellingen" (nr 6: Water voor Iedereen" en nr 7: "Energie voor Ontwikkeling" zijn van groot belang voor de ontwikkeling van kleinschalige waterkrachtcentrales, gelet op het belang voor lokale economische ontwikkeling en het opheffen van armoede van de lokale bevolking.

In dit onderzoek werd nagegaan of nieuwe versies van eenvoudige waterkrachtcentrales zoals watermolens een rol zouden kunnen spelen in met name ontwikkelingslanden. Er bestaat namelijk een toegenomen belangstelling voor kleinschalige waterkracht als duurzame energiebron in gebieden in Azië en Afrika waar nog beperkte energievoorzieningen aanwezig zijn.

Het EU-project HYLOW (2012) richtte zich op het ontwikkelen van waterkracht machines die kunnen werken met een klein verschil in waterspiegel. Het doel van het project was om twee types watermolens met een vrij oppervlak te onderzoeken, en een type in een gesloten leidingsysteem. Deze machines zouden een alternatief moeten bieden voor conventionele turbines in geval van een klein lokaal verval, en gebaseerd moeten zijn op milieuvriendelijke duurzame technieken. In dit proefschrift wordt een numeriek model ontwikkeld waarmee de vormgeving en eigenschappen van kleinschalige watermolens kunnen worden onderzocht. De aandacht ging met name uit naar het vinden van manieren om de efficiency van dergelijke apparaten te verbeteren.

Het gebruik van numerieke modellen om de prestaties van apparaten te beoordelen en voorspellen wordt in vele toepassingsgebieden gebruikt: in de luchtvaart, auto-industrie, en ook in de waterbouwkunde. Door steeds krachtiger computer hardware en software en is met name het gebruik van Computational Fluid Dynamics (CFD) sterk toegenomen voor het onderzoeken van complexe stromingssituaties inclusief het omgaan met vrij oppervlak en draaiende watermolens. Verschillende parameters zoals de keuze van het rekenrooster, randvoorwaarden, beginvoorwaarden, turbulentiemodellen etc. zijn in het kader van dit proefschrift onderzocht.

Om er zeker van te zijn dat de gekozen randvoorwaarden correct waren zijn de experimentele resultaten van een proefopstelling die in het HYLOW project zijn uitgevoerd gebruikt om het numerieke model te valideren. Gemeten stroomsnelheden en waterdiepten werden gebruikt als invoer voor het numerieke model. De berekende uitkomsten werden eerst geverifieerd en vervolgens gebruikt om de vormgeving en eigenschappen van het ontwerp verder te verbeteren. Door gebruik te maken van computervisualisatie kon het gedrag van de machine beter worden begrepen. Het basismodel bleek afdoende realistische uitkomsten te bieden die overeen kwamen met de metingen, waarna modificaties konden worden doorgerekend zoals veranderingen in invalshoek van de schoepen om het rendement verder te kunnen verbeteren.

In dit onderzoek is gebruik gemaakt van de commerciële CFD-code ANSYS–Fluent®. De resultaten gaven aan dat een verandering in invalshoek tot een hoger rendement kan leiden, althans voor de hier onderzochte gevallen. Op basis van visualisaties kon worden vastgesteld dat energieverliezen door turbulentie met name plaatsvinden rond de schoepen, waardoor het rendement afneemt. Om het effect van de breedte van het toegangskanaal te onderzoeken zijn simulaties uitgevoerd met verschillende verhoudingen in breedte van het waterrad ten opzichte van het toegangskanaal. Dit impliceerde dat nieuwe modellen moesten worden ontwikkeld op basis van vierhoekige elementen met plaatselijke verfijning ter plaatse van het schoepenrad. De resultaten lieten zien dat het rendement van de machine hiermee verder kan worden verbeterd. Op basis van deze berekeningen kon de optimale breedte-verhouding worden bepaald .

Aanpassingen aan de bodemeigenschappen bovenstrooms en benedenstrooms lieten zien dat ook deze invloed hebben en het rendement kunnen verbeteren. Experimenten met verbeterde vormgeving bij de aansluiting op de wanden gaven aan dat lekkage kon worden verminderd waardoor het rendement eveneens toenam, met name door aanpassingen aan de wanden benedenstrooms.

Het vermogen dat met deze kleinschalige waterkracht kan worden verkregen ligt in de orde van enige kilowatt, hetgeen het mogelijk maakt om deze te plaatsen in gebieden met slechts een beperkte infrastructuur. Met behulp van kleinschalige waterkrachtcentrales kan op die manier toch energie worden gewonnen die van groot belang kan zijn voor lokale economische ontwikkeling en het opheffen van armoede, overeenkomstig de doelstellingen van duurzame ontwikkeling zoals beoogd door de Verenigde Naties.

Contents

Nomenclature

Acronyms

1D - one-dimensional

2D - two-dimensional

3D - three-dimensional

CAD - Computer Aided Design

CFD - Computational Fluid Dynamics

CV-Control Volume

ESHA - European Small Hydropower Association

EU- European Union

EU-WFD – Water Framework Directive

FVM - Finite Volume Method

GW - gigawatts

MDGs – Millennium Development Goals

NGO - non-governmental organisation

NS - Navier-Stokes

PDE- Partial Differential Equations

RANS - Reynolds Averaged Navier-Stokes

RMS - Root Mean Square

rpm - revolutions per minute

SDGs - Sustainable Development Goals

TWh – terawatt hours

VoF - Volume of Fluid

UN - United Nations

UN-DESA Department of Economic and Social Affair

UNDP - United Nations Development Programme

UNESCO-United Nations Educational, Scientific and Cultural Organisation

Latin symbols

Symbol	Description	Dimension	SI Units
A	Area	L^2	m^2
a	acceleration	$L\,T^{-2}$	$m\,s^{-2}$
c	velocity (absolute velocity)	$L\,T^{-1}$	$m\,s^{-1}$
Cr	Courant number	-	-
D	diameter	L	m
F	force	$M\,L\,T^{-2}$	N
Fr	Froude number	-	-
g	gravitational constant	$L\,T^{-2}$	$9.81\ m\,s^{-2}$
H	total hydraulic head	L	m
Δh	head difference	L	m
h	head		
h_D	hydraulic diameter	L	m
L	characteristic length	L	m
M	torque	$M\,L^2\,T^{-2}$	Nm
m	mass	M	kg
$\dot{m}$	Mass flow	$M\,T^{-1}$	$kg\,s^{-1}$
n	rotational speed	T^{-1}	s^{-1}
P	power	$M\,L^2\,T^{-3}$	W
P_{mech}	mechanical power	$M\,L^2\,T^{-3}$	W
P_{theor}	theoretic available power	$M\,L^2\,T^{-3}$	W
p	pressure	$M\,L^{-2}$	$kg\,m^{-2}$
p_{stat}	static pressure	$M\,L$-2	$kg\ m$-²
Pe	Peclet number	-	-
Q	discharge	$M^3\,T^{-1}$	$m^3\,s^{-1}$
r	radius	L	m
Re	Reynolds number	-	-
t	time	T	s
Δt	time-step size	T	s
v	velocity	$L\,T^{-1}$	$m\,s^{-1}$
V	volume	L^3	m^3
W	Watt	$M\,L^2\,T^{-3}$	Nms^{-1}

Greek symbols

ρ	density	$M\,L^{-3}$	$kg\,m^{-3}$
μ	dynamic viscosity	$M\,L^{-1}\,T^{-1}$	$N\,s\,m^{-2}$
ι	kinematic viscosity	$L^2\,T^{-1}$	$m^2\,s^{-1}$
η	efficiency	-	-
ω	angular velocity	T^{-1}	$rad\,s^{-1}$

Chapter 1
Introduction

1.1 Energy demand

With a continued global population growth, a 70% growth in electricity demand is expected by the year 2035 (WWAP, 2014). The growing need for energy for a such rapidly increasing population is one of the most important challenges in the near future, as energy is vital for all other developmental processes. The impact of energy availability influences the living conditions of communities to a large extent. Without electricity there can be no water-purification, no health-care and no pharmaceuticals. Hence electricity is essential for life in the Twenty-first Century and an emphasis needs to be laid on power availability.

The use of renewable energies reduces the impact of greenhouse gases and other negative influences on the environment. Hydropower in its role in renewable energy sources is the largest contributor to this sector with the expected global share doubling, accounting for 30% of all electricity production by 2035 (UN-Water, 2015). Large hydropower projects require large investments and have to fulfil environmental standards. These are not always fulfilled, especially in countries with weak economies, due to the high costs involved.

Additionally, the costs for the planning and construction of the required infrastructure for an electricity grid further increase costs. The transfer of electricity to regions with low population densities is often not feasible. An alternative for such regions is local power generation. This can be achieved in various ways, with different environmental impacts. The installation of a diesel generator is one such example. Renewable energy sources in the form of photovoltaic, wind or small (low head) hydropower may be deployed where resources are available. The power generated may be connected to local users or be distributed within a local grid. Small hydropower using appropriate technology can be implemented on small scale projects. The tapping of hydropower resources with machines like water wheels can bring social benefits for the surrounding communities with local power generation. These social aspects are reflected in the United Nations Sustainable Development Goals (SDGs) which show the all-encompassing effects of energy availability. The SDGs are based on the success of the Millennium Goals and describe the challenges that lie ahead.

In the case of low head hydropower, various technologies are available. The prices of these technologies vary widely together with the requirements for the infrastructure. The engineering design depends on the hydraulic conditions at the site and the predicted energy harvest can be distributed to users. Here the minimum power requirements for a household or community need to be considered to match with the generation capacity of the unit. In rural areas with existing irrigation systems, the potential in terms of hydraulic head can be exploited using small units. These irrigation channels vary in size and discharge and a suitable machine design is required.

1.2 Hydropower

With an increase in world population, energy supply is becoming an increasingly important factor on a global scale. While more energy-efficient equipment is becoming popular in the industrial nations, the bulk of energy demands are carried by developing countries. As awareness and concern of global environmental issues like the burning of fossil fuels is growing, technologies are being developed to harvest renewable energy sources. Hence the significance of alternative sources as well as small scale energy production is increasing. Hydropower plays a significant role as an indigenous resource.

Modern large hydropower plants use turbines to generate electricity. Sites are chosen based on energy demand, hydrological, geological and environmental conditions of the region. Legal aspects and the social structure of the inhabitants of the region are also considered. Depending on the conditions at the power plant, different types of turbines are used. They are supplied with water stored in a reservoir with a dam or are constructed in rivers with considerable structural requirements. Fluctuations in energy production are small and the power generated is fed into centralised electricity grids. The environmental aspects of large hydropower plants are important as these structures block the natural flow of the river system. The incorporation of ecological solutions increase the costs of such projects and are often overlooked in the planning of such projects in developing countries.

In comparison with other renewable energy sources like wind and solar (photovoltaic), hydropower can generate uninterrupted energy. There are however temporal variations in river systems. These can be controlled in large hydropower schemes by reservoirs which can be used to meet peak generation or seasonal demand.

Hydropower schemes can be classified into different categories. They can be planned as single schemes which are used for only for power generation. Multi-purpose schemes cover a range of uses like power supply, irrigation, drinking water supply, drought and flood control, navigation or recreation. The schemes can be run-of-river or reservoirs with a dam. They can also be classified according to the hydraulic head (high, medium, low, very low) or by size (large, small or micro). Large systems are connected to major grids or national networks. Depending on the capacity they may also be isolated off-grid systems.

Large multi-purpose plants offer the advantage of multi-year regulation which reduces seasonal variation. These can be complemented with smaller single-purpose schemes as well as run-of-the-river plants to reduce ecological impact on the environment. Public opposition to large-scale projects contrary to rising energy prices in terms of growing demand requires an in-depth study of small scale project feasibility.

1.3 Low head hydropower – the HYLOW project

Low-head hydropower generation with very low discharge using water wheels or other devices which do not require reservoirs or dams across rivers have the advantage that they cause slight or no impact on the environment. They are termed as free-flow turbines and, using small scale power generators, do not require costly high-voltage transmission lines. They are an uninterrupted source of energy. This characteristic is of significance as a renewable energy source and cannot be replaced by other renewables like wind or solar energy. Information on free-flow turbine technology in the public domain is almost non-existent (Khan, Iqbal and Quaicoe, 2008).

Conventional low head hydropower is relatively expensive as the turbines have to be designed and manufactured according to individual specifications. Very low head hydropower generation using machines like water wheels comes as an alternative. These units can be adapted to serve remote communities as off-grid solutions.

The development of hydropower machines for very low fall heads was investigated in the EU project HYLOW "Hydropower converters for very low head differences" (HYLOW, 2012). The scope of the project was to develop two novel hydropower converters with free water surfaces and one in a closed system. These hydropower systems were to offer alternatives to conventional turbine technology in terms of environmental sustainability for very low head differences. The impact of the machines with free surfaces on the environment was estimated from different perspectives such as fish-mortality, sediment-transport and power generation. The development of appropriate technology was a part of the HYLOW project in which considerations for the implementation of the hydrostatic machines were investigated. The electricity generated by the machines was considered to drive various apparatus from basic lighting to power supply into the grid. Keeping in mind the fact that this technology was also intended for first-time users i.e. being implemented in remote areas where the local community is not familiar with electricity, the implementation of appropriate technology included various aspects like socio-economic impacts and health benefits of communities involved. The overall benefits of energy availability and its consequences on the living conditions is was considered. The manufacture of small hydropower machines for electricity generation may incur relatively high costs on a cost per kilowatt basis but brings with it other social and health benefits which cannot be assigned a monetary value. Rural communities lacking access to the electric grid in developing countries have relatively small electricity load requirements. Hence apt technology would mean a local modern energy system for isolated communities. This may consist of a centralised battery charging system and a mini-grid which can be powered by hydropower.

After the success of the United Nations Millennium Development Goals (MDGs) till 2015, the Sustainable Development Goals (SDGs) were set up for the period till 2030. The 17 goals are to promote development across the world. All the goals contain sustainability as a key factor for progress. Clean and affordable energy production has been included as a Goal #7 which has a direct influence on several other goals which rely on electricity, examples being clean water, health, education, economic growth, etc. Here low head hydropower can provide sustainable local energy production. This is of particular significance in parts of Asia and Africa where the infrastructure is weak. The potential for off-grid power generation is high, as large-scale hydropower plants and grid networks require considerable investment.

1.4 Hydrostatic pressure machine

The hydrostatic pressure machine developed in the EU's HYLOW project (HYLOW, 2012) is driven by the pressure difference in the flow at a low rotational speed. The machine runs at atmospheric pressure with free water surfaces on the upstream and downstream sides, thereby minimising the impact on fish. This also allows for sediment passage which reduces the environmental impact. The theory and initial model tests show high efficiencies for this segment of head difference. The machine is of simple construction, thereby making it cost effective.

The hydrostatic pressure machine was developed for small hydropower with a rating below 1000 kW. The machines operate with very low head up to 2.5m. Conventional turbines operating in this range are not cost effective. High-efficiency Kaplan turbines require high flow volumes at low head differences. As a result, these turbines have large diameters and require considerable civil works. The hydrostatic pressure machine operates in three locations: in rivers and at weirs, in free streams and in water supply infrastructure. For developing countries suitable modifications can be made. In regions with existing mill races or other channels such as irrigation channels, the implementation of this machine does not require extensive modifications of the existing hydraulic structures. As in all hydropower machines, the machine has to be designed to suit the flow conditions at the particular location. The machine modelled here is a free-flow machine based on the theory proposed in the thesis "Hydrostatic Pressure Converters for the Exploitation of Very Low Head Hydropower Potential" (Senior, 2007). The wheel is mounted in a channel with a fixed width. The machine consists of a rotating shaft and hub with blades mounted on it. The flat blades originate from the hub radially, blocking the channel when the blade tips reach the channel bed. On the upstream side, the hub is submerged while on the downstream side the water level reaches the hub.

The HYLOW project covers various aspects of low head hydropower generation with new technologies like the assessment of environment effects like sediment passage and fish passage, morphology and environmental impact. Scale models of the pressure converters were used to determine performance characteristics of the machine and to provide empirical coefficients for the theoretical models. Results of large-scale models tests and prototypes of the machines under laboratory conditions provide data for the consortium partners involved in the numerical modelling of the converters. The processes occurring within the flow can be explained through numerical models and which can monitor values such as velocity and pressure. Variations in geometry show the effects on the values for efficiency. The numerical modelling of the machine was a part of the project which covered a wide range of hydro-ecological aspects of a hydropower system. The physical models at the experimental setup provided the data for the verification of the numerical model. This numerical model can be used to explain the flow conditions for different discharges and to explain the processes occurring during operation. Modifications in the model can be used to optimise converter efficiencies as minor changes in physical models would result in an increase in terms of model construction and testing time. The turbine modelled here is meant for application in mill races, irrigation canals or in place of existing weirs.

1.5 Modelling approach

The numerical modelling of the hydrostatic pressure machine as presented in this thesis was done as a part of the HYLOW project. A physical scale model of the hydrostatic pressure machine was constructed and tested in the laboratories of the HYLOW partners at the University of Southampton and the Technische Universität Darmstadt. Tests were performed on the machine with various measurements being taken. The tests in the flume provided measured performance curves for the machine. These experiments enabled the development of an initial numerical model in a CFD environment based on the machine used in the flume. The model results were analysed and enabled a visualisation of the flow in the machine and its components. Modifications can be made to the model and the results analysed to examine the effects of the modifications and the effects on the efficiency of the machine. The high resolution of the mesh supplies detailed solutions for the problem posed.

The hydrostatic pressure machine and water wheels are low speed hydraulic machines which have a free water surface in front, inside and behind the wheel. The open channel upstream and downstream of the wheel has waves moving in different directions. Modern CFD codes have the option of modelling this interface between the phases air and water. Codes using the Volume of Fluid (VoF) method calculate the fractional volume of each phase within a cell. The free surface is then calculated with the fractional volumes of the cell and the surrounding cells. Some work has been done on simulation of slow-running hydropower machines using commercial Computational Fluid Dynamics software. The first CFD model of an ancient horizontal water mill (Pujol *et al.*, 2010) was run using the full computational domain. The model included modelling the two phases: water and air. Blade variations and their effect on performance using CFD on water wheels in the Himalayas showed results in agreement with experimental values obtained (Pujol *et al.*, 2015). Conventional hydropower turbines have a rotational symmetry. This allows for the reduction of the model size, which then consists of just one segment. The regions of the adjoining segments can be defined by symmetry planes leading to a considerable reduction of model size and computing time. In the machine modelled here a simplification is not possible due to the free surfaces.

With the increase in computational power, many codes are available for solving flow problems with CFD. Depending on the case, the codes can be adapted to solve the problem. Together with computer aided design (CAD) tools parametrised models with complex geometries can be modelled. Using these models, parameters within the CFD environment can be modified to observe the effects on the net result. In the case of rotating machinery, changes in rotational speed and its effects on the resulting torque and efficiency can be studied. The calculations are time dependant. Two-dimensional (2D) numerical models are a simplification of the case and provide a quantitative analysis of the case under study. The modelling is done with a unit model depth. This prevents representing cross flow during the simulation. This restriction enables an analysis of the processes in one plane during the operation of the wheel.

The model consists of a rotating wheel in a stationary channel. For the numerical model both the regions, or zones, are modelled separately and subsequently merged. In 2D, the geometry consists of lines and curves. The hub with blades are defined in the rotating zone with the channel bed, inlet and outlet are contained in the stationary zone. A spatial grid is created in the regions of each zone. The free surface is reconstructed in both zones for each time step. The resolution of the grid is determined by the equations used in the CFD solution method. For time-based problems, the dimensionless Courant number specifies the time step based on the velocity and grid size. Thus regions of high velocities require a finer mesh. These regions are found e.g. in the space between the rotating zone and the stationary zone.

The cases are defined by giving initial water levels and discharge. In the stationary zone, parts like inlets and outlets are defined by the purpose they serve. The rotating zone is given a rotational speed. For the simulations, parameters like inlet flow velocities specify mass flow or discharge. The boundary conditions describe the parts of the model and their function. The rotating wheel is defined by its rotational speed; inflow discharge is represented by the inlet condition, the outlet by a water level. The water surface separates the water phase in the channel from the air phase above it. The atmospheric pressure is defined in the air phase. For the simulations, the time step for each calculation is defined taking into consideration the expected velocities and the mesh refinement. The 2D model is set up for different water levels and rotational speeds and verified with experimental data. For the three-dimensional (3D) model of the machine, the numerical domain can be seen as an extension of the 2D model giving the wheel and channel a width. The wheel being the volume of the moving zone is completely enclosed by the stationary zone. In the 3D case, the VoF method reconstructs the free surface independently in each spatial dimension.

This work presents the approach to the numerical modelling of the machine and the parameters used in the setup of the simulation model. Various parameters such as the upstream and downstream water levels are discussed. Further development of the numerical model to improve design and efficiency of the machines will be presented. To ensure that the boundary conditions used are realistic, experimental results performed on a test rig were used to verify the models. With the help of this rig, flow measurements were performed around the machine. These measured quantities provide real boundary conditions for the numerical simulations. On the basis of the experimental data, the simulation results were checked and were used for the further development of the machines.

The conditions for the running of the simulation model with a range of discharges and various rotational speeds using different upstream and downstream water levels are shown to have a strong influence on the generated torque. Effects of the blade geometry can alter the pressure gradient in the flow and influence torque generation. Dimensions of the channel are altered for the flow to adapt better to the rotating turbine or wheel. The effects of altering the wheel width and narrowing the channel can change flow conditions within. A variation of these factors may change the performance of the machine. Different channels have varying slopes of the bed. These affect the flow and the effects of these can improve the conditions within the wheel sufficiently to improve efficiency. Similarly different channel dimensions on the upstream side and on the downstream side of the wheel may show some variations in the torque generated.

This approach enables a better understanding of flow conditions during operation through the visualisation of flow patterns. The basic model supplies results that are plausible and match those of the experiments enabling modifications of the machine geometry, e.g. blade angles, to be made on the simulation models. The model can then be so modified until a better efficiency is obtained in the simulations. In this context, this thesis examines the methodology required to develop and optimise very low head hydropower machines with free surfaces at a preliminary stage based on initial laboratory model data. The thesis also provides an analysis of very low head hydropower generation for machines with low rotational speeds and the infrastructure requirements for operation.

1.6 General Objectives

From an overall point of view, the scope of this thesis is to put into a global context, the chain of influence of power availability, for consumers, focussing on the development processes involved with the implementation of low head hydropower in developing countries. This is done by investigating the interaction of various factors which influence development in the field of power generation, its importance in developing countries, e.g. its importance for health, education, etc. The requirements for local power generation in rural areas in developing countries are explored. Therefore a numerical model of the hydropower machine to study the conditions within the flow will be developed. Based on this, the effects of modifications on the machine in its surrounding are to be examined. The potential of small hydropower will be explored from a global perspective.

1.7 Research Questions

The following will be addressed in this thesis:

1) What is the role of small hydropower in a global perspective?

2) Is a 2D analysis for water wheels sufficient for engineering design of machines or is a 3D analysis required?

3) What are the dominant parameters that need to be defined? What are the dominant parameters that influence the flow?

4) To what level of detail can the hydrodynamics of machines with free surfaces be simulated by numerical simulations now and in the near future? What role can CFD play in future developments?

5) How can small hydropower using appropriate technologies contribute to bringing about social change with the active participation of stakeholders?

1.8 Outline of the thesis

Chapter 1 introduces the scope of this research. This is followed by Chapter 2 which explores the role of energy availability with development. An approach for the implementation of technology in developing countries for power generation and the impact on local communities within the framework of the sustainable development goals is presented. The significance of small and micro hydropower in a rural setting and the influence on the population is discussed.

Chapter 3 presents an overview of hydropower machines and the machine to be modelled. The chapter introduces various hydropower machines, from the ancient water wheel to modern machines and the fundamental equations used to calculate the power in hydraulic systems. The machine developed in the HYLOW project is described.

Chapter 4 introduces the governing equations for CFD calculations and describes factors that play a role in the numerical modelling like discretisation in space and in time. Methods for modelling free water surfaces for simulating open channel flow are also presented.

Chapter 5 begins with the numerical modelling of a simplified 2D hydropower machine rotating in an open channel. The processes occurring during the functioning of the wheel are analysed, and the behaviour of the flow visualised. The complexity of the flow into the rotating wheel is presented along with the limitations of 2-dimensional modelling.

Chapter 6 focusses on 3D modelling of the machine. Flow conditions within the 3D domain are analysed. Variations of model size are studied and their impact on the results discussed. Alterations in blade geometry and their influence on the filling of the wheel segments upstream of the machine are explained. Asymmetrical blade configurations are implemented and their influences on the efficiency are studied. This chapter also considers the effects of geometry variations in the channel. The ratios of machine width to channel-width and the influence of other parameters in the channel on changes in efficiency are studied.

In Chapter 7 the results of CFD for small hydropower simulation is discussed, together with experimental observations. Environmental issues and economic potential of stand-alone power generation are addressed taking into consideration findings of the HYLOW project.

Chapter 8 presents the responses to the research questions with conclusions and lists the recommendations.

Chapter 2
The Water – Energy nexus

Economic progress critically depends on the availability of water. However, almost 1.3 billion people (20%) of the global population of 7 billion did not have access to electricity in 2010. With an expected 70% growth in the demand for electricity by 2035 (WWAP, 2014), the growing need for energy for a rapidly increasing population is one of the most important challenges, as energy is vital for all other developmental processes. The increase in energy demand will be covered by all types of primary energy sources. The impact of energy availability influences the living conditions of communities to a large extent. Without electricity there can be no water-purification, no health-care and no pharmaceuticals. Hence electricity is essential for life in the Twenty-first Century and an emphasis needs to be laid on power availability.

Hydropower is one of the oldest energy sources. In ancient times, water wheels were used to lift water or grind grain. In an early reference to water mills, the Greek poet Antipater describes the use of the machines to grind grain, thereby eliminating the need of young women to grind grain by hand. The earliest description of a water wheel is by the Roman engineer Vitruvius in the first century B.C. These water wheels were situated in or around settlements, providing a local power supply in the form of mechanical energy. The decline of hydropower began with the Industrial Revolution. Coal continued to be the primary energy carrier. The invention of the generator and of electricity enabled energy distribution. Coal became the primary energy source and a linking of power suppliers and consumers was enabled through a distribution network or electricity grid. The sheer size and number of coal power plants and the possibility to build them where energy was required gave this technology an edge over hydropower. The invention of the combustion engine fired by mineral oil increased overall power generation and enabled mobility. These decades of rapid advancement in industrialisation were accompanied by emissions resulting from the combustion of fossil fuels.

With the industrialisation in the Twentieth Century, the focus in energy turned on energy-intensive production technologies which are only feasible with large-scale units. The term "appropriate technology" was introduced in the context of developmental work to describe the socio-economic, political and ecological aspects of the technology applied. The environment where the technology is to be introduced is considered in terms of the choice of the technology (new inventions are seldom used), availability of local labour, staff and materials as well as the social and ecological impact of the technology to be introduced.

Concerns regarding the impact of technology on the environment as well as on society are primary issues in the Twenty–first Century. The various industrial revolutions since the invention of the steam engine have culminated in the awareness of finding solutions for technologies in use, to modify them in such a manner that future generations may be able to implement them without causing hazards to the environment. These hazards include air, land and water pollution, depletion of reserves in the form of ores as well as depletion of forests and pastoral land.

The driving force behind these factors is energy demand. With technological progress, modern conventional power generation plants run with improved efficiencies and reduced emissions. State-of –the-art technology is capable of producing zero-emission electricity. Coal was, and still remains the largest energy source thereby producing most of the carbon dioxide worldwide. Due to its attractive price, coal is the major energy carrier in most developing countries. The technologies often used have emission norms which are outdated. Affordability is the chief concern in small economies as funds are limited. The results of planned projects are that on completion, the units cannot produce sufficient electricity for growing demand through delays in the implementation of the power projects. In the world economy a continuing increase in efficiency in all fields has led to an increased use of resources, the end result being a larger gross consumption. Hence, in terms of sustainability, a limit ought to be imposed on the optimisation of efficiency (Grambow, 2013). In the last decades this has led to ozone depletion, global warming, reduction in biodiversity and other negative concerns. In the last decades these developments were recognised leading to counter measures being introduced under the title of "Sustainable Development". With the United Nations Millennium Development Goals (MDGs) set up in the year 2000 for the period till the end of 2015, the sustainable development goals (SDGs) will be the main focus till 2030. Based on the success of the eight MDGs targets, the SDGs were expanded to include seventeen detailed goals to promote development across the world. All the goals contain sustainability as a key factor for progress. Clean and affordable energy production has been included as a goal till 2030. This goal has a direct impact on many of the other goals which rely on electricity, examples being health, education, economic growth and industrialisation. Some other disciplines encompassed by the goals are climate, aquatic life, life on land and clean water and sanitation. Social aspects and sustainability are directly influenced by energy availability. The Sustainable Development Goal 7 is meant to ensure access to affordable, reliable, sustainable and modern energy for all (UNDP, 2015a). This goal emphasises on the potential of renewable energies as well as an energy mix. The efficiencies in the production of existing technology as well as new technology are to be increased. The goals are shown in Figure 2-1 below.

Figure 2-1: Sustainable Development Goals

Energy supply was not included in the Millennium Goals in spite of 1.3 billion people not having access to power. In Africa alone about half a billion people do not have access to electricity (Scheumann & Dombrowsky, 2014). In Africa, large hydropower projects are planned under the Africa-EU Energy Partnership to generate electricity in the megawatt range. This power will be consumed by users who have some experience of energy technology and where demand is high, but large hydro will not be able to cover rural areas comprehensively across the continent. Hence the SDGs will encompass a mix of energy systems which are to be developed for cogeneration. Hydropower here plays the role of being able to store energy through pump- storage plants thereby providing support for other energy systems like photovoltaic and wind power. The SDGs will also have to consider aspects which are drawn to by large infrastructure projects. Land acquisition, relocation of the population, ecological impact being a few. Further, the use of the power generated has to be distributed in such a manner that a maximum of people can access it. In a broader sense, the SDGs define access to various forms of energy. For regions which are connected to the grid, electricity is the main energy source. Should the household or settlement not possess connection to the grid, a local energy source like photovoltaic or wind as renewables or a diesel generator provide the necessary electricity. In the latter case the negative effects have to be taken into account. Here, the size of the grid is determined by the demand and the generating capacity.

The United Nations Development Programme (UNDP) led Multi-Functional Platform projects in West Africa have shown considerable success in improving the energy as well as overall contribution in several projects. The approach has been to consider energy production as the driving force behind progress. The impact of energy on the daily life of the community for example the access to water, can lead to gender equality as in many societies time-consuming tasks like fetching water or collecting firewood have traditionally been done by females. Energy availability (SDG Goal #7) could give them more time for schooling, lighting could enable them to study after dark. Through the introduction or electrical or mechanical devices to replace manual labour, more time may be available for meaningful work, thereby leading to gender equality (SDG Goal #5). The success of the UN's Millennium Development Goals in 15 years since 2000 shows that even though significant progress has been made in the drive to eradicate poverty, reduce child mortality, provide clean water and increase access to education (UN, 2015) there is still a lot of work to be done. Not all the goals have achieved their target. In 2015 the UN adopted the Sustainable Development Goals for the period till 2030, expanding the themes to cover 17 goals (UNDP, 2015b).

These goals aim to eradicate poverty and hunger. In a ranking of the SDGs for a developed country, Goal 13 "Take urgent action to combat climate change and its impacts" has the highest score followed by Goal 7 "Ensure access to affordable, reliable, sustainable, and modern energy for all" (Osborne, D Cutter, A Ullah, 2015). The goal rankings for developing countries would show a different order of priority, depending the current situation. In this chapter, the scenario for a country where Goal 7 has high priority is discussed, taking into consideration factors like poor infrastructure and remote locations. The implementation of small hydropower and the requirements for energy generation, distribution and its influence local communities is considered.

2.1 Large Hydropower

Hydropower schemes can be classified in different categories. They can be planned as single schemes or as multi-purpose schemes which are used for power generation, navigation, irrigation, water supply, flood control or recreation. The schemes can be run-of-river or reservoirs with a dam. They can also be classified according to the hydraulic head (high, medium, low, very low) or by size (large, small, micro or pico) hydropower. Large hydropower schemes produce power in the range of several gigawatts. These projects usually contain a number of turbines, but smaller projects may operate on just one turbine. The smaller systems can be connected to a grid or may be isolated off-grid systems. Sources for electrical power generation are shown in Figure 2-2. Hydropower accounts for almost 16% of the total electricity generation with other renewable sources contribution less than 5% of the total demand.

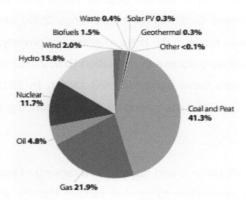

Figure 2-2: World electricity generation by source of energy as a percentage of world electricity generation in 2011

In comparison with other renewable energy sources like wind and solar (photovoltaic), hydropower can generate uninterrupted energy. There are however temporal variations in river systems. These can be controlled in large hydropower schemes by reservoirs which can be used to meet peak generation or seasonal demand. Multi-purpose schemes cover a range of uses like power supply, irrigation, drinking water supply, drought and flood control, and navigation. All hydropower schemes use the energy potential of flowing water to turn a turbine which converts the energy to mechanical energy. This drives a generator which produces energy which is then supplied to the grid in the form of electricity.

The main components of a large hydropower scheme are listed below:

- Dam: The dam across a river creates a reservoir which can store water.
- Intake structure: Water exits from the dam and is transported to the turbines through a pipeline (penstock)
- Turbine: The force of the water striking the turbine blades turns the turbine which is connected to a generator by a shaft.
- Generator: The rotating turbine turns the rotor of the generator. Magnets in the fixed-coil generator also rotate also producing an alternating electric current (AC).
- Transformer: The electric current is converted into a high-voltage current for effective transportation over long distances.
- Transmission lines: Transport the electricity to the grid. Transmission lines to and from remote areas are expensive.
- Outflow: The water exiting the turbine is transported back into the river through pipelines.

Large hydropower enables storage for dry periods by utilising reservoirs which can also influence flood control and navigation. These large schemes have their disadvantages and have drawbacks in their negative influence on the environment. The effect on regional biodiversity is also a negative factor. Similarly the subsequent social disruption and the loss of cultural and historical heritage plays a role in the execution of large-scale projects. The overall ecological impact of large hydropower is of significance. The worldwide hydroelectricity production was estimated to have increased by more than 5% in 2010 (WWAP, 2014). As shown in Figure 2-3, the growth rate of increased hydropower generation has kept up with that of other renewable energy sources combined.

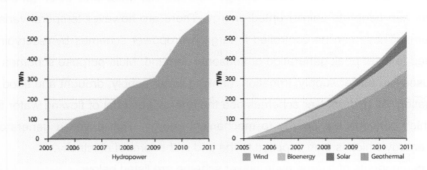

Figure 2-3: Increase in electricity generation from hydropower and other renewables

Hydropower potential in the world is considerable (Figure 2-4, Kumar et al. in (WWAP, 2014)). Economic feasibility shows that only an estimated two-thirds of the potential can be feasibly utilised. The gross share of hydropower in the energy mix is estimated to lie at 15% till 2035 (WWAP, 2014). The technical potential which is still underdeveloped begins with the African continent at 92%. This is followed by Asia (80%), Australia and Oceania (80%), and Latin America (74%). Here increases in power output are significant in emerging economies in Asia and South America with China and India and Brazil respectively. Social, economic as well as environmental issues predominate the implementation of large-scale projects.

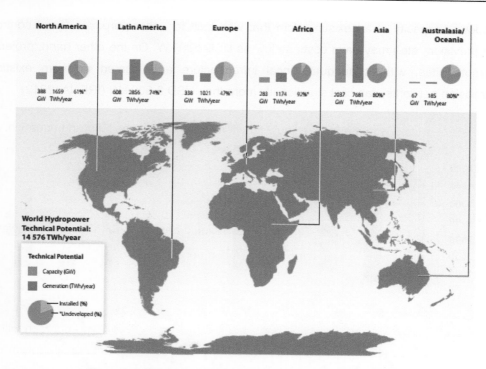

Figure 2-4: Worldwide technical hydropower potential in 2009

In the global energy mix the role of hydropower is significant. Large multi-purpose plants offer the advantage of multi-year regulation which reduces seasonal variation. These can be complemented with smaller single-purpose schemes as well as run-of-the-river plants to reduce ecological impact on the environment. Public opposition to large-scale projects contrary to rising energy prices in terms of growing demand requires an in-depth study of small scale project feasibility. Continent wise the emphasis can be on multi-year variation and implementing SHP schemes. Rapid economic growth in developing countries, especially in Africa needs to implement the role of low-carbon technologies in developing economies. Hydropower as an energy source is an important solution in ensuring sustainability as compared to other energy sources like fossil fuels. This resource can promote regional integration and eradicate poverty in terms of long-term clean energy supply in economic and social contexts. The total installed costs for large-scale hydropower projects typically range from a low of USD 1000/kW to around USD 3500/kW. However, it is not unusual to find projects with costs outside this range. For instance, installing hydropower capacity at an existing dam that was built for other purposes (flood control, water provision, etc.) may have costs as low as USD 500/kW. On the other hand, projects at remote sites, without adequate local infrastructure and located far from existing transmission networks, can cost significantly more than USD 3 500/kW (IRENA, 2012)

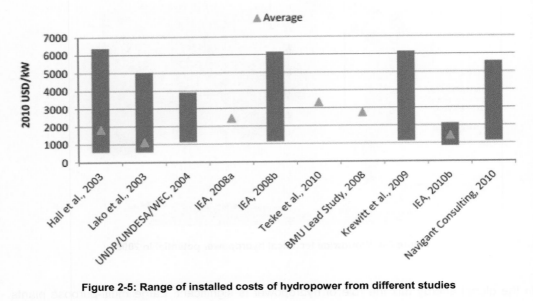

Figure 2-5: Range of installed costs of hydropower from different studies

2.2 Small Hydropower

Small hydropower schemes cannot be exactly defined worldwide. SHP, in general, describes plants which produce less than 10 MW of power. Mini and micro hydropower plants have an installed capacity of less than 500 kW and 100 kW respectively. The factors which determine this are discharge and head difference. As in the case of large hydropower plants, SHP has more flexibility in being connected to national grids or in local mini-grids along with other SHP plants. They can also operate as stand-alone units in remote areas. An additional advantage over large schemes is that SHP can be installed in conduit pipes and irrigation channels.

According to the European Small Hydropower Association (ESHA) and the European Union, SHP units are defined by a power generation capacity of less than 10 MW. This definition varies from country to country. With an increasing world population as well as advancing automation, the importance of SHP to cover energy demands in rural areas is increasing. Often large-scale projects are not feasible due to the high investments involved. Connecting remote villages in rural areas to the national grid is often more expensive than installing local generators and may involve deforestation at a large-scale. Developing countries often do not have the additional financial capacity required in the planning and building of infrastructure for complete electrification. Here low head hydropower could provide electricity for local consumption along river banks or irrigation channels. In the expanding economies of India and China, where a majority of the world population lives, units with a power rating of less than 25 MW are considered under SHP. With an estimated growth rate of 7-8% in India, the energy requirements will increase correspondingly (Bhat and Prakash, 2008). SHP has received much attention in recent years as a source of renewable energy as well as moderate investment costs to provide electricity in under developed regions. Considering the environmental impact, SHP does not involve major alterations to river courses which give it the distinct advantage of being able to be utilised locally or to be fed into the electricity grid. Today approximately 1.9 % of the world's total power capacity, 7 % of the total renewable energy capacity and 6.5 % (< 10 MW) of the total hydropower capacity is covered by SHP. This technology is at fifth place in terms of installed capacity to other renewable energy sources as shown in Figure 2-6 (UNIDO, 2016).

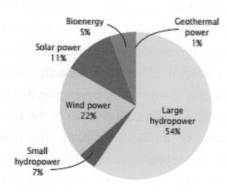

Figure 2-6: Global share of SHP

Different types of hydropower schemes exist in the SHP sector. Run-of the-river schemes utilise minimal water, and return temporarily diverted water to the water source whereby they do not require reservoirs. This enables more flexibility and ensures ecological compatibility with environmental norms. The potential for SHP plants worldwide is shown in Figure 2-7. Asia has the highest installed capacity while Europe is has developed the largest portion of 85% the resources in this segment. Other countries in larger continents have the potential to develop SHP in various schemes. Asian countries like Bhutan can integrate SHP projects in various land-use schemes including irrigation with power generation.

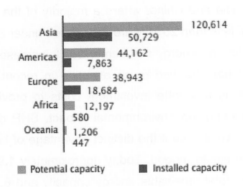

Figure 2-7: SHP below 10 MW capacity by region

Tributaries of big rivers can ensure a constant supply of hydroelectric energy in rural communities without a significant alteration of flow patterns of the river. These small-scale units can encourage rural communities to gain access to social medial services and promote growth in the region. In developing regions such as in sub-Saharan countries, small hydropower has a low level of installed capacity compared with the available potential. SHP can therefore play a key role in developing the potential into actual power needs for the countries considered. The Congo Basin or the Ethiopian Basin for example, can contribute to power supply on a nationwide basis. The potential of SHP in Africa is largely influenced by climatic and topographic conditions. The total SHP installed capacity for Africa is 580 MW and the total estimated potential is 12,197 MW. This indicates that approximately 5 % has so far been developed (UNIDO, 2016). SHP potential has been utilised and developed in Europe on a large-scale. According to a study by the European Renewable Energy Sources Transforming Our Region (RESTOR) some 50000 historical sites have been identified for potential hydropower generation. At these sites, the possibility exists to develop SHP units which contribute power into national grids. Regulation is a barrier which prevent development of SHP in many countries.

In the absence of reservoirs or dams small hydropower units which operate in isolated areas as off-grid or mini-grid schemes, flow variations over time need to be considered during the design and planning stages. Such plants have a rating of below 10MW. Conventional turbines like Kaplan, Francis and Pelton can be used in such installations. The main costs are for the turbine and piping framework since the construction of a dam is not required. Installation costs lie in the range of 2,000 to 4,000 Euro/kW (ESHA 2013 in (Michalena and Hills, 2013)). Drawbacks of the development of SHP schemes in remote areas without infrastructure are the absence of roads. Infrastructure-building schemes need to be implemented to provide accessability to these regions. For larger installations transmission lines are required. Often financial support from local banks is absent. Operation and maintenance need trained staff which requires capacity building.

2.3 Appropriate Technology

Hydropower in all its diversity uses different technologies for different working conditions. The machines being developed are being considered to drive various apparatus from basic lighting to power supply into the grid. Keeping in mind the fact that this technology being developed is also intended for first-time users i.e. being implemented in remote areas where the local community is not familiar with electricity, this study on appropriate technology includes various aspects like socio-economic impacts and health benefits of communities involved. A holistic approach to the implementation of the technology is considered in order to estimate the total benefits of the users. Technology that is conceived for small scale applications and local consumption can be termed appropriate if it has a minimal impact on the environment. The labour intensive applications of appropriate technology are decentralised and the energy efficiency is high. The technologies can also be appropriate in terms of environmental, cultural, social, ethical, political or economic perspectives. Its applications can be found in developing countries or under-developed areas of industrialised countries. In developing countries this technology often fills the gap left by conventional urban development which involves high costs. The technology addresses sustainability and thus minimizes environmental impact.

Renewable energy, also called alternative energy, consists of sustainable energy sources. These energy sources (renewables), seen from a human time scale, are continually available and are naturally replenished, in contrast to fossil and nuclear fuels which are being depleted. Physically energy can neither be created nor destroyed. The term renewables, in general, refers to systems where energy is extracted from natural process and then used in technical applications. These systems extract energy from eco-systems thus intervening in the eco-systems themselves. Such sources are sunlight, wind, water, tides and geothermal heat. In a recent legislation, the European Union (EU) has passed a directive on the promotion of renewables with the objective of generating 20% of the power requirements using renewable energy sources within the EU by the year 2020 (European Parliament, 2009). Modern high-performance electronics have raised the efficiency of energy generation from renewable energy sources in the form of electricity. Low voltage photovoltaic power generation has become possible due to advances in inverter technology (conversion of DC to AC). This technology can be applied to very low head hydropower to enable the operation of standard 220 Volt (50 Hz) electrical devices such as light bulbs, refrigerators, etc.

Some characteristics of renewable energy sources are given below.
- Solar: Photovoltaic technology has made rapid progress in the past few years and serves as an alternative for off-grid rural households. The technology has been implemented in remote areas and power can be generated as 12 V DC current as well as 220V AC with the help of an inverter. The panels are available commercially.
- Wind: Available wind mills involve high investment costs and require large installation as wind speed increases with the distance from the ground. The construction of such towers requires considerable infrastructure. Wind energy generation has become widespread in off-shore wind farms.

- Hydro: As mentioned in the previous sub-chapter, large hydropower plants are expensive in financial terms as well as their impact on the environment. These plants are connected to large-scale grids. Conventional low head hydropower is relatively expensive as the turbines have to be designed and manufactured according to individual specifications. Very low head hydropower generation using water wheels comes as an alternative as the costs involved are considerably lower than turbines. These units can be adapted to serve remote communities.

Implementation and impact

Any power generation technology requires some maintenance. These costs are looked at as a percentage of investment costs on an annual basis. Longer maintenance-free periods indicate lower operating costs. Robust technology may be more expensive but enables a smoother functioning of the unit in the long run. The manufacture of small water wheels for electricity generation may incur relatively high costs on a cost per kilowatt basis but brings with it other social and health benefits which cannot be assigned a monetary value. Rural communities lacking access to the electric grid in developing countries have relatively small electricity load requirements. This makes transmission and distribution of grid services less attractive for local providers in developing countries. Hence apt technology would mean a local modern energy system for isolated communities. This may consist of a centralised battery charging system and a mini-grid which can be powered by hydropower alone or hybridised with wind or photovoltaic systems. The overall potential of the introduction of such units has to be seen over a longer period. The benefits of availability of power are direct as well as indirect. Figure 2-8 shows some of the uses. Direct benefits e.g. water supply which can be enabled by pumping water into a tank. Indirect benefits, like the improvement of health standards within the community, are made possible by storing medical supplies etc. in a better manner (refrigeration).

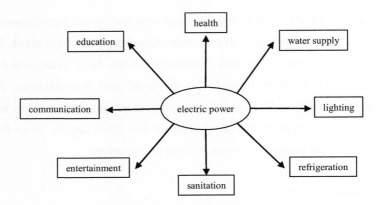

Figure 2-8: Benefits of electricity in society

The different aspects pertaining to the topics in the figure will be handled in relation to the hydropower machines being developed in the subsequent chapters.

Significance of local hydropower generation

The programme of international research and collaboration on a New Path or World Development presented to the UNESCO Natural Sciences Sector Retreat identifies five areas of interest to improve living conditions globally (Erdelen, 2009). The underlying causes for today's crises can be categorised in three groups: climate, resources, environment; poverty and global development; and the global economic and financial situation. The New Path combines a systematic focus on five areas of interest, "clusters" of connected issues within an integrated programme and considers the complexity and interconnectedness of critical international issues within a coherent systems framework (Erdelen, 2009). The five "clusters" are:

- Climate, Environment and Resources
- Economic Restructuring, Finance and Globalisation
- International Development
- Social Transformation
- Peace and Security

The challenges projected in the field of international development include an estimated increase of world population of 2.3 billion people within the next forty years. Together with the 2 billion people existing on less than 2$ per day insufficient to cover their basic needs of health food and nutrition, poverty, political insecurity and environmental degradation are inevitable. A doubling of the global economy in the next 20 years would imply that an additional 2 billion people will have living standards of today's middle class with the same patterns of consumption and waste. A population rise from 6.7 to 9 billion will result in an unsustainable use of biological and other resources. Food production cannot meet the demands and instability and migration will be the result. The increasing stress on water resources will be aggravated through overuse, mismanagement and contamination in addition to the effects of climate change. The effects of the latter through desertification will intensify competition for limited resources.

The inequality in the distribution of opportunity, income and wealth show that the present trend of global economic development is unsustainable. Hence new policies and ideas for economic growth and a better distribution of wealth and opportunity worldwide are necessary. These policies have to keep environmental requirements in mind. Substantial flows of finance will be needed to support the change in policy towards a path of equitable and sustainable world development (Erdelen, 2009). These finances have to include the real costs for the environment, social and human costs as well as the real costs for resources and energy.

Hence the need for appropriate and sustainable technology arises. Technological progress is the key to growth in a modern economy. Advances in science and technology increase the knowledge and skills of workers. The restructuring of economies to suit sustainability can therefore be achieved by laying emphasis on science and education. Development can be thus brought about by the dissemination and application of suitable technologies. Keeping in mind the conservation of, for example water resources, energy and efficiency, the existing technologies may be adapted and modified to meet specific needs and objectives. Emphasis should be laid on the spread of new renewable energies with very low emission levels. As capacities in developing countries are weak in terms of building new sustainable societies, technologies introduced in such regions should have some institutional characteristics which would enable the indigenous societies to improve themselves. They should be able to use the potential of the technology and knowledge to understand that the conservation of the environment, employment and equitable development are interlinked and should be given priority. Education and public information are of critical importance as they can alter cultural values of society and the individual.

Eco-efficiency analysis

Economic and ecological sustainability can be related through an eco-efficiency analysis. The global interconnectedness of production processes imposes that measures taken to improve one environmental aspect have effects on other environmental aspects (Huppes, 2007). As economic growth is usually larger than environmental improvement, the environment is forced to deteriorate. Deterioration of the environment is inevitable if no measures are taken to improve the environment along with economic growth. The value created through economic activity and the relationship to the environment, or the balance between these two issues are not viewed from the social aspects which have to be dealt with separately. Appropriate technology and sustainable development can be supported by eco-efficiency analysis which gives an insight into what can be expected to occur between the economic and environmental aspects, providing a link between the values created and the society directly affected. The analysis can also be focussed on the costs thereby showing the environmental cost-effectiveness or environmental improvement per unit of cost (Huppes, 2007).

Social aspects

The aim of providing renewable lighting to households would replace the use of paraffin or other substance and would offer an important environmental benefit by reducing greenhouse gas emissions at the rural community level. More importantly the reduction of hazardous gases formed as a result of burning fuel will improve indoor air quality health conditions of inhabitants. The improved indoor lighting can enable adults as well as children to study better at night. If alterations in the watercourse are to be made, this can result in a better water storage facility during the summer months and reduce the toil of women and children to transport the water to their homes. Storage of water also enables better crop irrigation. In addition to these effects, the work within the community could spill over to other community efforts.

Some of these aspects can be seen in pilot projects in hill tracts of the Western Ghats of India. Three such pilot projects have been carried out in the state of Kerala, India initiated by the UNDP's Small Grants Programme (Ebrahimian, 2003). The three remote villages not connected to the grid are small clusters of a few houses each spread over a large area. The main source of livelihood is agriculture in gardens around the households. Some animal husbandry is also often present. The main source of lighting is kerosene and firewood is used for cooking. Usually womenfolk walk considerable distances to collect water. In a collective effort, the whole community combined efforts and installed turbines to generate electricity. The community runs and maintains the hydropower plant and generates 16 kW (peak) of electricity. During summer months, the community relies on a diesel generator. Community members pay for electricity consumed. The efficiency of the project is partly due to the fact that the whole community is literate.

The technology implemented will, in general, have to be subsidised as the rural dweller in most isolated communities in developing regions cannot afford high investment costs. Further, a suitable local person has to be employed with the aim of contacting the authorities in case of a breakdown or malfunction. Here technical assistance will be required to solve the problem. Spare parts like bulbs/LED will have to be available locally as the inhabitants cannot be expected to travel for long distances to obtain them. Hence a suitable strategy has to be planned to facilitate sustainable development with appropriate technology in developing countries. In a funded photovoltaic project to electrify rural communities in Botswana, studies revealed that over 80% of selected households disconnected the photovoltaic systems after two years due to lack of technical support (Ketlogetswe, 2009). Here a process-based approach may be considered favourable. The building and operating of a water wheel power generator by the community itself together with the maintenance can generate an interest in the preservation of the local forests.

Remote areas in developing countries are rarely visited. The indigenous population can be characterised by poverty and illiteracy. The introduction of power supply can be linked to a general improvement of living conditions. This could enable a better utilisation of time (water pumps may reduce the time-consuming fetching of water) and can be supplemented with learning programs. The social dimension[1] of electrification of villages and communities and development projects is intricately linked to technology. Thus "technology in a social context" approach considers all affected parties, from members of the public, engineering designers as well as decision makers as part of the social context, thus seeing in them stakeholders or potential stakeholders (Russel, A.W. Vanclay, 2007).

Sustainable energy systems

The lack of sustainable development in the energy sector increases the risks of climate change, resource wars and nuclear accidents (Hennicke, 2005). The implementation of sustainable technologies in the micro power-generation sector in developing countries as well as in industrialised countries offers a large potential in terms of electrification as well as reduction of greenhouse gases etc. It combines the participation of rural population in partaking of their share of energy with the conservation and protection of resources, the environment, health and climate. The cost-effectiveness of sustainable technology in the long run is to deliver least cost energy. The management of decentralised diversified energy systems has to be enabled through robust and efficient technology. The introduction of environmentally viable crops and new patterns of sustainable consumption and production can help in improving the standard of living further. By enhancing energy productivity, economic growth can be decoupled from non-renewable energy consumption.

This aspect has to be taken into consideration along with the design criteria. The power generation capacity of the unit has to consider the number of households, livelihood and the social activity and the available energy source. Were modifications e.g. in the channel to be made, sustainability can be ensured if the hydro plant is multipurpose i.e. irrigation or other social activity (AHEC, 2005). Employment increasing measures can be introduced such as powering handlooms etc. to give local craftsmen a foundation for economic improvement.

---

[1] The meaning of 'social dimension' relates to things that affect people's way of life, their culture, their community, their political systems, their environment, their health and wellbeing, their personal and property rights, and their fears and aspirations (Vanclay, 2003)

Power grids vs. stand-alone units

Power generated at a power plant can be used at the location of the plant itself, or can be fed into a power grid which distributes the electricity over a larger area. As mentioned earlier for large plants like those built at dams, the amount of power generated is so high that it cannot be utilised locally. Industrialised nations in Europe use international grids to meet demands on a continental scale. Such structures are not available in developing countries as they often do not have enough power to meet their own demands. In many of these countries the national grids are in a bad state of repair due to lack of maintenance and outdated technology. Power supply by private companies and organisations was not as widespread as it is today, with many countries recognising the need to reduce carbon emissions and use renewable energy sources. Off grid technology is a quite recent option and is still being developed. High performance electronics has made rapid progress in the inverter sector, enabling the conversion of electricity form to suit that of the grid. In general, hydropower generation capacities of 5 kW and above are fed into a grid. The maintenance of the complete system requires a considerable infrastructure starting from the turbines, grid synchronisation, load control and transmission till it reaches the electricity meter of the consumer. Generation units below 5 kW are better suited for local consumption, the recipients often being communities or small groups of communities. Figure 2-9 shows the possibilities in bringing power to the end user. The machine modelled in this thesis lies on the lower scale and is thus suitable for isolated generation. The maximum power output of the machine considered lies at around 10 kW, with the option of feeding the electricity into the grid.

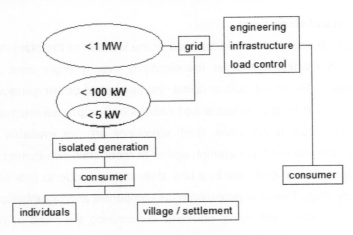

Figure 2-9: Power generation

Grid generation requires that the power generated is in the form of an alternating current in which the flow of the current changes direction periodically. The rotational energy at the shaft of a turbine is converted into electricity using either a synchronous or an asynchronous generator. The synchronous generator is a machine which has the ability to control current and reactive power It is a complex and expensive machine which requires considerable maintenance. Additionally the machine requires grid synchronisation. The asynchronous machine is a simple, robust, low maintenance machine which is cost-effective and enables easy synchronisation with the grid. The reactive power of the machine is compensated by capacitors. The generator can run at variable speeds but the total efficiency of the machine is reduced by a complicated inverter.

In spite of the considerable costs of asynchronous generators, this solution is held to be appropriate for the machines being developed. The generators offered are available as kits and can be purchased once the specifications of the machines are known. This eliminated the need for complex construction plans for the electrical parts of the hydropower machines. As each machine is a unique model for a particular location, the generator can be acquired after completion of the basic design. It is well known that the generation of electricity to be supplied into the grid (e.g. 220 V, 50 Hz) from slow-moving machine shafts supplying high torque is a major challenge. Present day windmills face the same challenge. Historically, gear boxes have been mainly designed to reduce the rotational speed of engine shafts. The reverse process requires extensive design skills which specialist companies possess. Standard electro-mechanical machinery is available in major cities in developing countries. However the costs are high as the equipment is often not produced locally e.g. most generators and motors available in Ethiopia are of Chinese manufacture. The costs are high as there are hardly any local manufacturers to compete with these products in terms of quality.

In general, with markets opening worldwide, basic materials like steel, motors etc. are available worldwide. Due to different import duty structures, the prices of materials not produced locally are higher. Key components like ball-bearings and electro-mechanical machinery required for the machines need to be of high quality to ensure smooth functioning over a long period. This will result in higher initial costs. The situation in many Asian developing countries like India is improving. The infrastructure, which has been developing here at a slow pace, is improving rapidly due to the improvement of manufacturing facilities as well as due to foreign investments. These foreign investments not only bring along technology but also additional benefits like the training of local staff. This process has been going on for some decades and has resulted in many indigenous manufacturers being able to produce machines like 3-phase asynchronous spur gear motors of a high quality. Additionally, change in policy is also encouraging local production of machinery.

Grid generation demands high investment costs. From the commercial aspect, the break-even time should acceptable to ensure economic viability. Hence capital as well as operating expenses are high which make the technology suitable for developed regions. This technology is suitable for regions where skilled personnel are available for maintenance. The complexity of machinery is high, as it involves backup systems to prevent large damages in the case of breakdown. Off-grid hydro needs to store the energy harnessed. This is done using batteries which can store low voltage (up to 48 V) DC. Over and under charging protection can be installed using a shunt regulator. Most systems use deep cycle batteries which are designed to have 1000 to 2000 cycles at 80% charge (5 to 15 years). Further, machines operating at low voltages need thick cabling to reduce power loss which increases the costs of the units.

This shows the need for a threshold in the machine's generating capacity for supplying power into the grid. For low power generation capacity up to about 1 kW, the power can be utilised locally. This eliminates the need for the asynchronous generator which can be replaced by a DC generator. Here too, the best solution is to avail of ready-made kits which suit the specifications of the machines. Figure 2-10 shows the set up for generating power with small hydropower. The generated DC current can charge batteries via a charge controller or be fed into an inverter to drive the household electrical load. The machine may be situated in a water channel/river nearby.

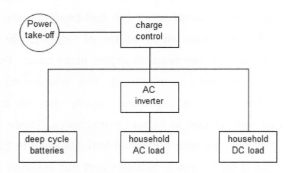

Figure 2-10: Power generation, storage and conversion

The basic electricity requirements for a household in a rural area are listed in Table 2-1 together with the operational times. Power which is generated during periods where demand is very low can be stored in batteries. This could occur during the night when only the refrigerator is in operation together with a lamp. Based on these figures for one household and given the number of households to be provided with electricity, the feasibility and rating of the machine may be determined given the hydrology on the water source.

The minimum requirements should enable some surplus power for future expansion. In the case, the planning of a television has been included to provide for extra power required when the community has covered their basic needs and has found some means of additional income e.g. through effective income generation programmes.

Table 2-1: Basic power requirements for a household

Device	Number	Daily operation	Power per unit [W]
Lamps	5	7	60
Refrigerator	1	24	100
Pumping	1	2	120
Radio/TV	1	6	30

Based on the load pattern for a typical household (Figure 2-11), a good estimation of the required generation capacity and the potential for storage can be obtained. The sum of the requirements will also give the dimensions of an alternate energy source required if the water course is not perennial.

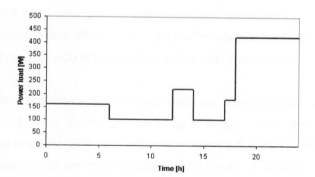

Figure 2-11: Electrical load pattern over 24 hours

The design of the converter is governed by the hydrology at the location. Depending on flow charts over a series flow measurements over large periods of time give an estimation of the design of flow of the machines. The dimensions of the machine are based on the design flow.

Infrastructure requirements

Details of the machines being developed can be found in the relevant reports on the construction and design of the prototypes. As in the case of large hydropower, the generating system consists of three groups of components:

- civil works: diversion works, channels, piping
- power generation equipment: turbines, water wheel, drive systems linking the turbine or wheel to a generator and/or mechanical device, a generator, turbine and generator controller, switchgear and transformer
- power distribution system: power distribution lines and consumer service connections

A basic water wheel consists of a shaft and a hub with attached blades. The shaft can be connected to a generator (electricity) or to a mechanical device (mill to grind grain). The diameter and breadth of the wheel depend on the existing structures (channel or weir) at the location and have to be designed individually. Additionally, these dimensions depend on the available fall head (difference between the upstream and downstream water level).

Large, slow-moving water masses as often found in old water mills together with large wheels generate a high torque which is good for mechanical work. Electricity generation requires high speeds to match the frequency of the grid. This dilemma can be solved with modern inverter technology for various applications such as grid-connected and off-grid systems. Due to their high costs it would be advantageous to run DC appliances for off-grid solutions.

The list of common parts can be split into three groups:
Non-movable parts at the channel

- trash rack (upstream of the machine)
- flow inlet structure
- wheel mounts (gabions) in wheel chamber
- wheel bearings
- flow outlet structure

Movable parts

- trap to shut inlet
- wheel
- wheel shaft
- bearings

Power take-off

- clutch
- gearbox
- electro-mechanical generator unit
- electronic load controller
- wiring
- storage batteries
- inverter
- control panel

As an example, the hydrostatic pressure machine from the HYLOW project is described below, pointing out its salient features and fields of application. The hydropower machine is submerged up to its hub thereby functioning like a weir to maintain the upstream water level. Figure 2-12 shows the machine in the flume from the downstream side. As the pressure of the water acting on the hub and blades is large, civil structures are required to anchor the machine.

Figure 2-12: Laboratory model (view from downstream)

The power output range can begin at a higher rating (ca. 2 kW) making the power take-off suitable for grid generation. The safety measures for this machine are considerable and require an elaborate construction as the machine acts as a weir. Additionally, if the power generated is to be fed into the grid, adequate measures have to be taken to ensure that the electro-mechanical equipment does not get damaged in the case of a major breakdown during generation. Infrastructure is required to assemble the machine and to mount the machine. As high Voltages are expected at the generator, trained personnel are required to maintain the unit. As the parts required for construction of the machine increase with the size of the machine, local availability may hinder the implementation of these machines in many countries. Again, the dimensioning of the machine depends on the hydrology at the particular location. The machines can be installed in irrigation channels, in streams or at locations with existing weirs whereby local legal requirements need to be considered during the planning stage. In irrigation channels, the concerned authorities may have to authorise the installations.

Areas of application of hydropower machinery

The following Table 2-2 shows regions and locations for the application of the machine. Apart from the vital parts like ball bearings, shafts, generator etc. required, the table also shows possible variations in the material used for other parts.

Table 2-2: Areas of application of the machine

Region	Location	Materials	Generation	Uses
rural	rivers irrigation channels weirs	wheel: metal/steel	off grid DC off grid AC grid	electrical: lighting refrigeration storage
urban	mill races weirs	blades: metal/steel	off grid AC	mechanical: grinding milling handloom
developed	small dams dikes	mounting: concrete	grid AC	water level regulation

Hybrid systems - cogeneration

In comparison to other renewable energy sources like photovoltaic solar panels which has overall efficiencies of 0.2 to 0.3, small hydropower at around 0.7 is of significance on a global scale. Cogeneration and hybrid power systems are based on combinations of various systems like wind turbines with photovoltaic panels or photovoltaic with small hydro. Here multiple technologies are integrated in a single system for power generation. The overall efficiency of the system is the product of the individual systems. The advantage of this technology is that an uninterrupted power supply can be obtained over a 24-hour period in comparison to other stand-alone renewables like photovoltaic or wind energy systems. Seasonal fluctuations in water level and flow have to be taken into account especially in countries where wet seasons like monsoon occur on an annual basis. The machine can also function as a weir providing overflow while generating power simultaneously. As opposed to photovoltaic technology, the capacity of these machines cannot be increased. Hence considerations during dimensioning of the machine have to include foreseeable increase in power demand in the future. An aspect to be considered is the integration of water wheels in other renewable energy systems like photovoltaic. Here additional power output can be obtained when demand exists. As mentioned earlier, here too technical support has to be provided to avoid disconnection of the system within a short period of time as in the case studied in Botswana.

2.4 Small scale power generation

Local power demand

The choice of particular hydropower machines depends on the targeted community. The resources and requirements of remote communities are limited. The applications in terms of power demand are usually limited to lighting and refrigeration. As pointed out earlier, power requirements in regions without any electricity have low demand, typically in the range of 200-300 watts. Infrastructure is generally not available. Applications in such regions would be autonomous generating units meant for local consumption. This fact will simplify the power take-off as the generated electricity does not need to meet the specifications/requirements of the grid (frequency). The larger the flow requirements are, the more sensitive the machine becomes to fluctuations in flow. To counter this, elaborate measures have to be taken to ensure a steady flow as well as to guarantee the minimal flow in the water channel or river. This can only be achieved with a good maintenance by trained staff. The hydropower machines under construction are capable of supplying up to 10 kW. This is the peak capacity and will vary according to flow conditions. In tropical countries the power generation varies on an annual cycle making additional back-up systems necessary. As a reference, the power to run a 60W light bulb could be generated using a wheel with a radius of 0.5 m running at a rotational speed of 7.5 rpm. DC power could be taken off a dynamo running on the rim of the wheel. This would enable the unit to supply power locally. Similarly, given a head of 1 m and a discharge of 1 m³/s would generate 7.3 kW of power theoretically. A higher wattage also implies greater investment costs and more safety considerations. Easy access to the power unit has to be guaranteed as in the case of a breakdown, the extent of influence of the loss in power is considerable. Areas where local power consumption is vital e.g. to provide refrigeration for medical supplies in rural areas in the tropics, the reliability of the machine has to be guaranteed.

Developing countries with growing populations are candidates for the application of modern renewable energies. In most cases the existing power grids are overloaded (in terms of consumers) at peak periods. Many of these countries generate power for their base load with outdated thermal power stations. Poorly functioning public administration as well as bad power management further aggravates the situation. Small hydropower technology can contribute to carbon-free power generation by providing local communities with energy. It is estimated that in the Himalayas, stretching from Myanmar in the east to Afghanistan in the west, an estimated 500,000 watermills exist (ITPI, 2007). Considering the potential of about 200,000 watermills in India (Vashisht, 2012), the commercial aspect of electricity generation with water wheels in the region can help improve the living standards of local communities. Further, the utilisation of existing water mills to generate mechanical energy has to be considered. Many remote villages in the hills often sustain themselves on agriculture. Most types of grain have to be mechanically handled in some form or the other. Rice has to be de-husked, wheat ground. This tedious work is usually carried out by women themselves, or it has to be transported over long distances to be milled and subsequently transported back. Studies have shown that the benefits of rural electrification with RE have improved the overall situation of the communities living in remote villages where no access roads exist (Mohapatra, G. Ali, A. Mukherjee, 2009). By introducing machines to mill grain, people from the surroundings will be encouraged to use the opportunity to grind their grain locally.

Government policies on local power generation

India has enabled small scale power generation by changing policies. Previously, in what was considered a rigid structure with a few governmental institutions having the power generation monopoly, it was virtually impossible to generate power in the small and mini hydro sector. These conditions have drastically changed with the legislation on the renewable energy act with the target of generating 20% of its energy requirements by 2020 from the renewable energy sector. The amendments in the act make it possible for persons to set up "captive generating plants" which permit "any person to generate electricity primarily for his own use and includes a power plant set up by any co-operative society or association of persons for generating electricity primarily for use of members of such cooperative society or association" (Ministry of Law and Justice, 2003). Thus the act emphasises on rural electrification with the focus on enabling local cooperatives and governing bodies to license free generation and distribution of electricity. This also enables the efforts of non-governmental organisations (NGO) to integrate power generation in their programs. Similarly the Government of China is promoting small hydropower generation in a significant manner. By the end of 2015, the total hydropower capacity of China reached 320 GW with an annual output of 1,100 TWh. (UNIDO, 2016). Approximately 80 countries around the world have introduced local or national feed-in-tariffs and other incentives to make power generation with renewable energy sources more attractive for investors.

2.5 The HYLOW hydrostatic pressure converters

It is estimated that an unused small hydropower potential of 5 GW exists in Europe. The range of these hydropower resources lies between 50 and 1000 kW. This hydropower I is mainly available at existing weirs, with the River Thames e.g. having a potential of 75 MW (head differences from 1.4 to 2.4 m) alone. For the UK, the potential lies between 600 to

1000 MW; for Germany around 500 MW is available (HYLOW, 2012). In the HYLOW project, the applications three new turbines developed for small hydropower can be more cost effective than conventional technology. The new technology has a minimal impact on the environment. The hydrostatic pressure converter was meant for rivers and existing weirs. The second machine was meant to be applied in free stream situations and the third machine for small head differences in existing water distribution networks. A simplified variation of the hydrostatic pressure machine was also considered. The machines operate at very low head differences of below 2.5 m. Conventional technology is not suitable for very low heads as the costs involved are very high. Implementation of Kaplan turbines will require large civil engineering works and the diameter of the turbines will be considerable. The costs per kW for a Kaplan turbine lie at about 7,500 Euro with an additional 7,500 Euro per kW for the civil structures. The ecological consequences are high as river continuity is disrupted and the movement of fish is hampered. Fluvial ecosystems may also be endangered. The intention of the project was to develop small hydropower turbines for the head range up to 2.5 m. The machines were to be cost effective and were thought for installations up to 1000 kW. Environmental impact was to be kept at a minimum. The areas of application of the turbines were in free surface flows in rivers and channels at weirs, in free stream flows and in water pipeline networks with low heads. To realise this, the machines were investigated from theoretically aspects and by using and physically models. Tests were performed on the physical and numerical models developed leading to the optimisation of the machines. Prototype installations were used to evaluate the ecological impact in situ, enabling a comparison with scaled models tested in the laboratory.

In view of the current energy demand, the adaptation of the machines for application in various scenarios in developed and in developing countries were considered favourable. The technology was developed in the project using theoretical and physical models. Based on the theory, the optimisation of the machine was done with the physical and numerical models. The large-scale models consist of two prototypes with power ratings of 10kW and 5 kW. Of the two installations for the prototypes, the larger machine was situated in Bulgaria with the other being in Germany. The larger machine was integrated in a weir without the energy being fed into the grid. The smaller unit was connected to the grid.

2.6 Environmental issues

As a source of green and clean energy, hydropower contributes significantly to the world power demands. Several countries are handling hydropower development on a priority basis. In the case of large hydropower plants with dams and reservoirs the impact on ecosystems is far reaching. The fragmentation of river systems brings economic and developmental advantages and benefits human activity. However, the construction of dams affect flow regimes by altering the seasonal variation of flows and eliminates seasonal flooding of floodplains and wetlands thereby affecting river ecology. As a consequence the breeding grounds of aquatic species are affected. The levels of sediments transported downstream is reduced which in turn affects coastal ecosystems which are supported by healthy sediment flows. For electricity generation with regulated flows, factors which need to be considered are water and erosion regulation on the upstream side. The direct impacts are on fish and aquatic life, the downstream water flow regulation and sediment transport.

Simple hydropower machines like the converters developed in the HYLOW project have the potential to provide urban and rural/isolated areas with a reliable, efficient, safe and economic source of energy, in decentralised areas and industries. Their relative simplicity allows for good environmental impact as they have slow rotational speeds and do not completely block the water course.

The machines for low head hydropower are seen to have a lower impact on the environment as compared to conventional turbines. They do not break the continuity of the flow and do not block the passage for sediment of plants and fish. As the hydraulic head is small, no large pressure differences occur which may be harmful to aquatic life. The flow rates at the outlet match with those of the inlet, fish are not attracted to this region as the flow at fish passes is higher. The impact of the converters on the environment was demonstrated at the installations. The main issues addressed by the HYLOW project partners are described below.

Morphodynamics

The relevant project partner investigated sediment erosion and deposition for the hydropower converters. The machines allow for sediment passage and physical model testing in the flume showed the movement of trapped sediment and sediment deposition Analysis and documentation of river bed topography at run-of-river installations were performed. Model tests were performed on a HPM installation at a weir to observe the machines influence on mobilisation, transport and deposition of sediments. An analysis of erosion patterns was made and the introduction of countermeasures for negative patterns was developed in the form of guidance structures.

Fish response to the machines

The impact of the floating converter on the local environment was assessed along with impact on fish in a laboratory model. Observations of the impact on fish in a natural habitat were studied with regard to residential and migratory species. Prediction models on aquatic ecosystems were developed. Fish mortality and prediction of mechanical damage to the fish in the form of blade strike while moving downstream was studied using numerical models. Assessment of the probability of blade strike based on a prototype of the converter in variance of discharge, operation and species of fish was performed. The influence of the model on geomorphological processes and the impact of these processes on invertebrates were studied. The impacts of the machines in hindrance of migration routes, delays in migration, and the subsequent increase in energy requirements and predation risks were investigated. The influence of the machine on the hydrodynamics and the subsequent impact of vortices in the flow and vortex generation on fish behaviour was also studied upstream and downstream of the machine. The introduction of a fish pass especially for upstream migration together with downstream migration was considered. An environmental impact statement for the hydropower converters could be derived from an analysis of the results.

Water Framework Directive

The European Water Framework Directive legislation ensures e.g. the continuity of rivers to improve the ecology of streams and river systems. The blocking of rivers using dams and weirs to create the head for hydropower generation interrupts the water courses. This has a negative impact on the ecology of the system. The Directive addresses legal framework regarding development of hydropower in rivers by assessing the technology being introduced. The compliance of technology through an assessment with the regulations along with permission requirements on regulatory bodies applies in particular to the EU. These processes will also have to be initiated in the country of implementation of power generation. Slow moving machines like the hydrostatic pressure machine permit fish passage and the transport of sediment in the river which are also considerations in the Directive. Mechanical damage to fish is minimal and the free water surfaces do not create large pressure differences. Sediment can pass through the machines as bed load and suspended load. The Directive also states that it is necessary for water bodies and water courses throughout EU to be characterised as having a "good ecological" status or, if modified largely, have a "good ecological potential". To reach these commitments degradation in ecological quality is banned and any alterations must lead to the protection, enhancement or restoration of the water body. The installations of the HYLOW hydrostatic pressure converters were planned within the framework of the Directive. Discussions with the concerned authorities in the early stages of the planning could ensure that the installations constructed were in fulfilled the requirements stipulated in the Directive.

2.7 Costs

With no current technology available for very low head hydropower conversion, many remote regions are without electricity. For irrigation channel networks some electricity is needed for measurement equipment or other equipment. Water reservoirs are often located in remote areas without access to the grid. Substantial adjustment can be made on the units to accommodate ecological factors as well as provide satisfactory technical solutions for on-site implementation. The efficiency of local power generation is not influenced by transmission losses over long distances. The availability of untapped water resources in developing countries can contribute to carbon-free energy and an overall improvement of communities in remote areas as in the case of small hydropower generation discussed earlier. Efficiencies for conventional turbines like the Kaplan are high, but the installations0% but require considerable civil works. They also require large diameters for high discharges at low heads. A detailed comparison of hydropower machines utilising low hydraulic heads up to 2.5m is given in a study by Bozhinova et al., 2012 (Table 2-3). The following gives an overview of low head hydropower generation technologies in terms of efficiencies, costs and environmental considerations. The HYLOW hydrostatic pressure machines showed low cost, low damage to fish and the possibility of sediment passage. Costs for the hydrostatic pressure wheel were very low.

Table 2-3: Comparison of parameters for low head machines

Type	H: m	Q_{Design}[a]: m³/s	P: kW	Mechanical efficiency: %	Costs: L/M/H	Fish damage: L/M/H	Sediment passage: Y/N
Impulse wheels	0·4–1·5	1·0–8·0	1·4–45·0[b]	35–40	L	M	Y
Poncelet wheel	0·7–1·7	0·3–6·0	1·0–55·0[b]	55–65	M	H	Y
Zuppinger wheel	0·7–1·5	0·7–6·0	5·0–60·0[a]	70–75	M	L	Y
Francis turbine	0·75–5·0	1·0–10·0	10·0–200·0	75–85	M	M	N
Kaplan turbine	1·8–5·0	1·0–25·0	10·0–1800·0	82–92	H	H	N
VLH	1·4–3·2	10·0–30·0	100·0–500·0	80–86	L–M	L	N
Archimedes screw	1·0–10·0	0·1–5·5	1·0–300·0	up to 80	M	L	N
Vortex converter	0·5–2·5	0·5–20·0	1·0–200·0	41	M	L	Y
HPM	1·0–2·5	1·0–5·0 (≈20·0?)	7·5–50·0 (≈240·0)[b]	70–82	L	L	Y
HPW	0·2–1·0	0·5–10·0	1·0–75·0[b]	60–90	very L	L	L

[a]Design flow rate for an energy converter with maximum geometric dimensions
[b]Assuming a width of 5 m
H, high; HPM, hydrostatic pressure machine; HPW, hydrostatic pressure wheel; L, low; M, medium; N, no; Y, yes

Similar to small hydropower installations, decentralised energy production, reduces the dependence of central power stations and transmissions. The low-cost machines can be installed in rural areas with low, medium and small head sites where power consumption is significantly low thereby increasing living standards. For the HYLOW project, cost estimations were difficult as such power generations had not been built. At project begin, the converters existed only as theoretical models and test models in laboratories. As the low head differences are uneconomical, an assumed German feed in rate of 9.67 €/kWh for a break-even period of 15 years was taken. This gave a cost estimation for turbines up to 500 kW at 7.250 €/kW. For larger installations 1000 kW (feed in rate of 6.5 €c/kW) the costs dropped to 4,875 €/kW installed capacity. The former amount corresponded approximately to prices for an English 200 kW power station (2005) and IWPDC information (2005) where 8,000 €/kW was given. Cost estimated for the HYLOW large-scale model, the machine were expected to lie in the range of 2,500 to 3,500 €/kW. (ESHA estimates showed the range of 1,250 to 3,500 €/kW for SHP in Europe.) The HYLOW Large-scale Model evaluation showed the prices per kW installed capacity.

Chapter 3
Hydropower take–off mechanics

3.1 Hydraulic machines

3.1.1 Turbines

Turbines are machines that extract energy from a flow and convert it into mechanical energy. The mechanical energy is converted into electrical energy which is distributed through a grid. In hydropower turbines the flow at the blades causes the turbine to rotate. The turbine shaft is connected to a generator to produce electricity which is then transported via the grid. Turbines can be classified into two categories: (i) reaction turbines and (ii) impulse turbines. Reaction turbines function on the principle of a pressure difference in the flow before and after the turbine. Impulse turbines are driven by the velocity of the flow. The machines operate under atmospheric pressure. A special form of a reaction turbine is a pump-turbine. This is a dual-purpose machine which generates power when water from an upstream reservoir flows through it to a downstream basin. When the electricity grid has an excess of power the turbine operates as a pump, transporting the water from the basin to the upstream reservoir. Depending on the head difference and the discharge, different types of turbines are used. For very high heads Pelton turbines are preferred. Water wheels are used for very low heads and discharges. Figure 3-1 (Giesecke, Heimerl and Mosonyi, 2014) shows the range in which different types of turbines are deployed. The mechanical efficiencies of modern turbines lie at over 90%.

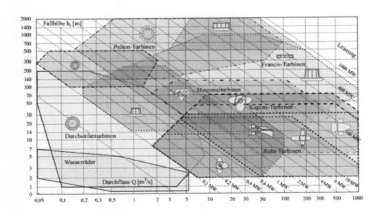

Figure 3-1: Deployment of hydropower machines

Conventional turbines often have complex intake structures to manage sediment transport. In the case of the damming of a river, the continuity of the flow is interrupted which has an impact on the ecological system of the river. These are corrected by structures like sand traps on the upstream side and fish-ladders connecting the upstream and downstream sides.

Kaplan Turbine

The Kaplan turbine is a reaction turbine. The pressure components within the flow are only partially converted into velocity components, thereby creating a pressure gradient within the runner. The pressure difference between the upstream side and the downstream side of the turbine is compensated by a draft tube on the downstream side of the turbine. Reaction turbines can be constructed with vertical or horizontal axes, or with inclined axes (Figure 3-2, a). The Kaplan turbine can cover a wide range of head differences at various rotational speeds. Another variation is the bulb turbine. The turbine has a horizontal shaft and is enclosed in a bulb-like structure together with the generator unit. The efficiency of bulb turbines is higher than that of Kaplan turbines as the direction of the flow does not have to be corrected.

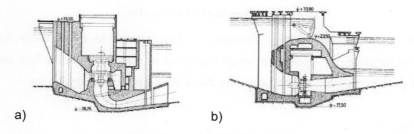

a)   b)

Figure 3-2: Kaplan (a) and Straflo (b) t urbines

Variations of the Kaplan turbine have evolved based on the flow rate and mounting. Beginning with Kaplan turbines with movable runner blades, the turbine has been further developed in the form of pit turbine, compact axial turbine like the Straight-Flow turbine (Straflo turbine) with the generator aligned concentrically around the turbine housing (Figure 3-2 b)) (Miller and Escher Wyss, 1974). More advanced concepts include double-regulated propellers to increase performance in small power stations. The HYDROMATRIX is a combination of more units with horizontal axes. Each small unit consists of a propeller turbine with a generator. In the case of run-of-the-river schemes, several of these units can be arranged across the river.

Francis Turbine

The Francis turbine is a reaction turbine. The turbine has a spiral casing around it thereby enabling the flow to enter the runner from all sides (Figure 3-3) (Grote, K.-H.; Feldhusen, 2007). Guide vanes at the inlet direct the flow onto the blades of the runner. The flow enters the machine in radial direction and exits in axial direction towards the draft tube. The turbines can have a vertical or a horizontal axis. They can be deployed with a rating in the range of 10kW to 750 MW. They can also be deployed in power plants with head differences over 700m.

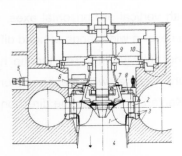

Figure 3-3`: Section of a Francis turbine

Due to higher specific speeds, the diameters of Francis turbines are smaller for high discharges. The turbines can operate at small variations in discharge as the flow can only be regulated by the guide vanes. The flow exits the machine through a draft tube.

Pelton Turbine

Pelton turbines are impulse turbines in which water exits jets and strike cup-shaped buckets mounted on the circumference of the runner (Figure 3-4) (Sigloch, 2013). The impulse of the water velocity exiting the jets and striking the buckets creates a torque and causes the runner to turn. The kinetic energy remaining in the water exiting the buckets is very low. The turbine wheel operates at atmospheric pressure and the pressure head in the flow is converted completely into kinetic energy on exiting the jets.

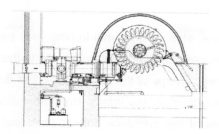

Figure 3-4: Pelton Turbine with horizontal axis

Pelton turbines are installed in power plants having high head differences and low discharges. The turbines can have a horizontal or a vertical axis.

The Very-Low-Head (VLH) Turbine

The VLH turbine has beeen developed for head differences in the range 1.5 to 3.0 m and consists of a regulated propeller-generator unit. The machine has a large diameter and is mounted in a simple square channel. The housing can be raised and lowered and the plades are protected by a trash rack. Depending on the head difference, the machine can be mounted at a 45° inclination to the channel bed (Figure 3-5) (Juhrig, 2013).

Figure 3-5: Very low head turbine in channel

Several units can be placed next to each other, similar to the HYDROMATRIX array. The turbine can be run for flow rates ranging from 9 to 27 m³/s per unit (MJ2 Technologies, 2016).

Cross Flow Turbines

In 1933 the cross flow turbine, also called Banki-turbine, was patented by Ossberger. The turbine functions on the principle of an impulse turbine. As in the case of water wheels, the machine is driven by the kinetic energy of the flow. It is a cross flow machine in which a free water jet enters the machine through the intake structure (situated either horizontally or vertically), strikes the blades of the cylindrical runner (impulse) and exits the machine in free fall (Figure 3-6) (Ossberger GmbH + Co, 2015). The width of the runner can be changed to match the flow.

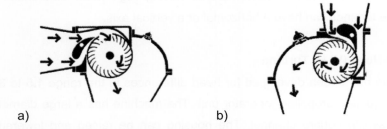

a)   b)

Figure 3-6: Cross flow turbine with (a) horizontal inlet and (b) vertical inlet

The free fall of the flow exiting the machine can be influenced by constructing a diffusor to control the downstream water level as is done in the case of conventional hydropower machines. The range of application of cross flow turbines lies between 2 m and 200 m for a wide range of discharges (Figure 3-7) (Ossberger GmbH + Co, 2015). These values depend on the manufacturer's specifications. For strong fluctuations in discharge the total channel width may be divided in two or three units operating in individual cells.

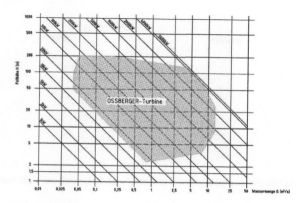

Figure 3-7: Application range of Ossberger cross flow turbine

The above figure shows that Ossberger turbines can be deployed over a range of head differences and discharges.

3.1.2 Waterwheels

Water-powered machines (waterwheels) have developed over centuries. In the course of this development, variations of each type came about. Water mills are still used in various parts of the world to grind grain. Water wheels are still used for power generation in many parts of the developed countries. Today the majority of the hydropower machines installed generate electricity across a wide range of ratings from a few kilowatts to some 100 kilowatts. Evidence shows that the waterwheel was in use in southern Mesopotamia around 3000 B.C. and was seen as a cultural object as well as a tool and a machine (Giesecke, J; Mosonyi, 1997). The first hydraulic machines were water wheels. In his writings in around 20 B.C. Vitruvius mentions the use of hydropower to rotate millstones with vertical axes using undershot waterwheels (Figure 3-8) (Pollio, V; Frontinus, 1528). The wheels themselves had a horizontal axis and were driven by the kinetic energy of the flow. The energy conversion to mechanical form led to various uses e.g. grinding grain, hammering, pounding, sawing and so on.

Figure 3-8: Vitruvius concept of a waterwheel with vertical axis millstone

Waterwheels have been in operation since the early middle ages till well into the Twentieth Century with wood being the main construction material till about 1800. With the beginning of the Industrial Revolution, the significance of waterwheels began to decline. Over the course of centuries, the simple water wheel has undergone numerous changes. Ferdinand Redtenbacher describes a water wheel as a wheel-shaped hydraulic machine with receptacles at the circumference on which the water imparts pressure or an impulse (Melorose, Perroy and Careas, 2015). In his analysis of the existing waterwheel technology of the time, he emphasises on the progress made by the development of the Poncelet wheel (Figure 3-9) (Melorose, Perroy and Careas, 2015). While existing waterwheels were driven by the momentum of the flow or by the pressure of the flow or, in parts both, the Poncelet wheel was designed to allow the water to enter the wheel smoothly while maintaining a constant pressure on the blades. The blades are curved to enable water to enter in direction of the circumference without impulse, and exit with a low velocity. Water enters through a sluice and, after striking the blades, exits directly as the blades are not closed on their inner diameter.

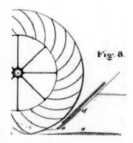

Figure 3-9: Redtenbacher's drawing of a Poncelet wheel

In general, vertical waterwheels may be classified as overshot, breastshot or undershot wheels. All three types generate a high torque and have a low rotational speed. The wheels are usually fed by a runnel which carries the water to the waterwheel. The three types are shown in Figure 3-10 (a,), (b) and (c) respectively (Bach, 1886).

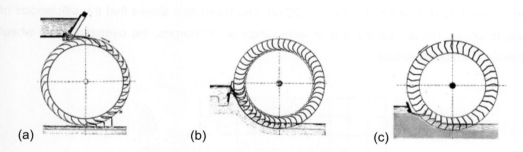

(a)   (b)   (c)

Figure 3-10: The overshot (a), breastshot (b), undershot (c) waterwheels

Breastshot waterwheels are fed by water at the height of the centre of the wheel or above it. Undershot waterwheels are rotated by water striking blades below the wheel. The introduction of curved blades improved efficiencies of vertical machines like the Poncelet wheel. Further developments in blade geometry led to the construction of a wheel by Zuppinger. The change in the curvature of the blades enables a better filling and emptying of the chambers thereby improving efficiency. Figure 3-11 shows the water wheel developed by Zuppinger (Bach, 1886).

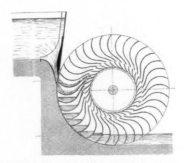

Figure 3-11: Zuppinger wheel

Though water wheels with vertical axes existed, they are not mentioned in the textbooks of the Nineteenth Century, the main reason being the rapid advances in the development of turbines. Hence theory and construction guidelines only applied to vertical wheels with horizontal axes. The efficiency of water wheels in comparison to modern turbines is shown in Figure 3-12 (Giesecke, J; Mosonyi, 2009). The figure also shows that the efficiencies of water wheels remain steady over a wide range of discharges, the overshot water wheel having the highest values.

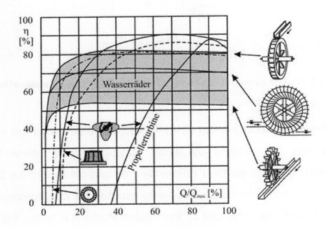

Figure 3-12: Efficiencies of waterwheels and turbines

The transition from waterwheel to the first machine to use the reactionary forces of the flow was invented in the Eighteenth Century by Johann Andreas von Segner. Machines with vertical axes were developed by Fourneyron and Henschel in the Nineteenth Century. Through the introduction of turbine technology in the middle of the Nineteenth Century, machines with vertical axes gained significance and began to replace horizontal axis waterwheels. Today, water wheels are still used for low head differences and discharges. The machines are robust and require low-maintenance. Installations are in the range of up to about 100kW.

3.1.3 Other hydropower machines

Hydrodynamic screw

The hydrodynamic screw is an ancient machine, earlier used to lift water from rivers and other water bodies to irrigate fields. The machine consists of helical surfaces attached to a shaft (axis). The helical surfaces act as blades. The shaft is set at an incline between the upper and lower water surfaces. Figure 3-13 shows an Archimedes screw (Giesecke, Heimerl and Mosonyi, 2014). The whole apparatus is set in a cylindrical chamber. The machine is attributed to Archimedes of Syracuse and is also known as the Archimedes' screw. It was operated manually by rotating the shaft and surfaces. In the Netherlands the screws were driven by windmills to drain polders. Modern screws are used to generate electricity or to pump water. In water treatment plants screws are used to transport sludge.

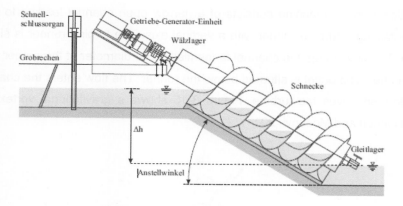

Figure 3-13: Hydrodynamic screw

Hydropower screws are available commercially. The screw operates at heads between 1 and 10 metres for flows between 0.2 and 15 m³/s (Landustrie, 2016). The operating speed of the screw is low which makes it less dangerous for fish. The machine requires a stable foundation to hold the trough of the screw. Areas of application are at new or existing weirs or channels with a head drop. The maximum power rating lies around 1000kW as shown in Figure 3-14 (Landustrie, 2016).

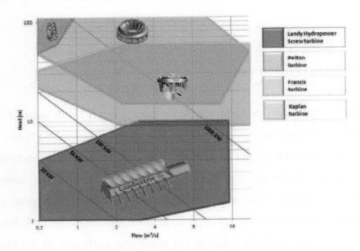

Figure 3-14: Application range for hydropower screws

Gravitational Vortex Machine

The gravitational vortex machine consists of a narrow open channel leading to a circular chamber containing a turbine runner with a vertical axis. The turbine runner is similar to a water wheel. The flow enters the channel tangentially. The diameter of the runner is smaller than the chamber. The outlet is situated below the runner. The flow enters the chamber and the vortex formed drives the turbine. Figure 3-15 shows a gravitational vortex machine (Giesecke, Heimerl and Mosonyi, 2014).

Figure 3-15: Gravitational vortex machine

The application range of the gravitational vortex turbine lies between 0.7 and 2.0 m head for discharges between 0.02 and 20 m³/s (Zotlöterer, 2016). Due to low rotational speeds, a gearbox and an inverter are required to feed electricity into the grid.

Free flow turbines

Research is being done on various other hydroelectric power generation machines for low head hydropower. These free-flow turbines include horizontal axial turbines as well as cross flow turbines such as the Darrius and Savonius type as shown in the following Figure 3-16 (Khan, Iqbal and Quaicoe, 2008).

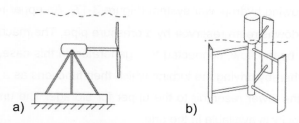

a)   b)

Figure 3-16: Free-flow turbines: a) axial turbine, b) cross flow Darrieus turbine

These free-flow machines are mainly propeller type turbines or derivations. Most of the machines involve underwater mountings involving increasing costs and are considered to have a negative ecological impact e.g. through blade strike from turbines.

3.1.4 Power calculation

The theoretical attainable power P (in Watt) contained in a hydraulic system can be calculated using the system schematisation as sketched in Figure 3-17. For a head difference of Δh and a discharge Q, not considering losses in the system, the theoretical maximum power is expressed by

$$P_{theor} = \rho \cdot g \cdot Q \cdot \Delta h \; [W]$$

3-1

as elaborated by (Giesecke, J; Mosonyi, 2009). This power potential for a head difference of Δh is shown in the following hydropower system (Figure 3-17). An upper reservoir with a dam is connected to the downstream reservoir by a pressure pipe. The machine installed is a turbine which is driven by the flow, connected to a generator. In this case, the generator can be used as a motor thereby driving the turbine which then functions as a pump, thereby transporting water from the lower reservoir to the upper for storage. The unit is run in this mode when excess electricity is available in the grid.

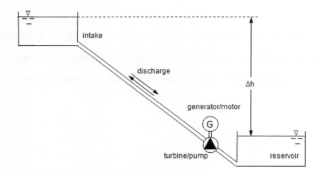

Figure 3-17: Hydraulic potential in a pump-turbine system

The efficiency of the complete hydropower system is influenced by hydraulic losses due to pipe friction, blade design, leakage, etc. The total efficiency of the hydraulic system is determined by the product of all the efficiencies of the individual components. For rotating machines within the flow, the turning moment M_T acting on the turbine is given by the Euler turbine equation which describes the blade geometry of conventional hydropower turbines

(Giesecke, J; Mosonyi, 2009) This equation describes, as in Newton's Second. that the turning moment or the turbine is equal to the reactionary force $-M_s$, which in turn depends on the flow situation in the turbine. The perimetral inflow and outflow velocities $\bar{c}$ vary along the radius r of the blades.

$$M_T = -M_S = \rho \cdot Q \cdot (\bar{c}_{2u} \cdot r_2 - \bar{c}_{3u} \cdot r_3) \ [Nm]$$

3-2

using the following notation:

M_T	turning moment on turbine shaft due to flow
M_S	reaction moment of boundaries (blades)
Q	discharge
$\bar{c}_{2u}$	perimetral inflow velocity into the control volume
$\bar{c}_{3u}$	perimetral outflow velocity from the control volume
r_2	radius at inlet
r_3	radius at outlet

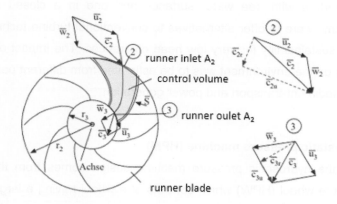

Figure 3-18: Velocity triangles at turbine runner

The equation uses the absolute flow velocities $\bar{c}_i$ which are calculated from the relative meridional (fluid) velocities $\bar{w}_i$ and the circumferential velocity $\bar{u}_i$. Figure 3-18 shows the relationship between the velocity triangles and the blade geometry (Giesecke, J; Mosonyi, 2009). The theoretical power P_{theo} obtained at the turbine runner is the product of the turning moment M_T and the angular velocity ω of the runner given by equation 3-3.

$$P_{theo} = \omega \cdot M_T \ [W]$$
                                                                                                3-3

The efficiency of a turbine can be expressed in terms of how close the results of equation 3-3 match those of equation 3-1. These equations are used to calculate the blade geometries for conventional hydropower plants which use the potential energy of water stored in reservoirs to drive turbines.

3.2 The HYLOW project

3.2.1 Project scope

The development of hydropower machines for very low fall heads was investigated in the EU HYLOW project (HYLOW, 2012). The scope of the project was to develop two novel hydropower converters with free water surfaces and one in a closed system. These hydropower systems were to offer alternatives to conventional turbine technology in terms of environmental sustainability for very low head differences. The impact of the machines with free surfaces on the environment was to be estimated from different perspectives such as fish-mortality, sediment-transport and power generation.

3.2.2 The hydrostatic pressure machine (HPM)

The principle of the hydrostatic pressure machine design comes from the conventional hydrostatic pressure wheel (HPW) which consists of a wheel having a large diameter and straight blades emanating radially from the axis. The height of the blade is greater than the height of the upstream water column (Figure 3-19) (EU-FP7, 2007). The pressure difference between the upstream and downstream water column drives the blades.

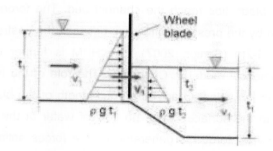

Figure 3-19: Hydrostatic pressure wheel

The wheel is partially submerged on the downstream side. There is a step in the channel bed below the axis of the wheel. In contrast, the hydrostatic pressure machine has a large hub with blades emanating radially. The water level can reach up to (or over) the hub (Figure 3-20), thereby reducing the blade length. As in the case of the hydrostatic pressure wheel, there is a drop in the channel bed below the machine. The machine can be adapted to be installed in weirs in existing channels.

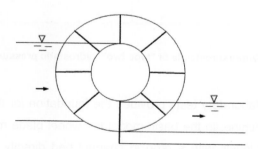

Figure 3-20: Hydrostatic pressure machine principle

For the HPM, the upstream bed is slightly higher with a curved segment of the bed running along the circumference of the wheel. The bed has a step below the axis of the wheel leading to a lower downstream side bed.

The numerical modelling of the HPM in this thesis is based on the machine described in the dissertation of Senior as a "type two" hydrostatic pressure convertor (Senior, 2007). The wheel is mounted in a channel with a fixed width. The machine consists of a rotating shaft and hub with blades mounted on it. The flat blades originate from the hub radially, blocking the channel when the blade tips reach the channel bed. The force acting on the blade below the hub is driven by the pressure difference between the upstream and downstream water column (Figure 3-21) (Senior, 2007). The HPM is further described during the construction of the numerical model in Chapter 5. The role of the hub is important as it permits the pressure on the wheel to act on the full length of the blade. For the hub, the pressure acts normal to the surface. Since no flow of water at the hub takes place, no resultant work is done. The torque is generated by the forces acting on the submerged blades.

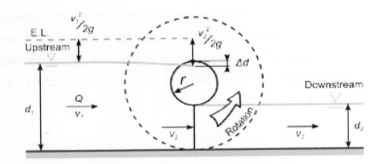

Figure 3-21: Working principle of "type two" hydrostatic pressure converter

In the HYLOW project, laboratory tests provided the foundation for the development of the hydrostatic pressure machine. In the basic model the wheel blade maintains the upstream and downstream water levels with a drop in channel bed directly below the axis of the wheel. The pressure difference between upstream and downstream water levels drives the wheel as shown in Figure 3-19 (EU-FP7, 2007). Advantages of these types of machines are that the velocity of the flow is maintained which allows for a larger discharge to pass through the machine. It also maintains riverbed continuity and river continuity which enables the migration of aquatic life and sediment transportation. Low rotational speeds reduce the ecological impact. The machines can be installed in existing structures like mill races and weirs. No complex control systems are required which indicates that the machines are cost-effective and can be installed as stand-alone units in remote areas. Initial installations in the laboratories of the HYLOW project partners show that the machines can be used for energy production effectively in ecological an economic terms.

3.2.3 The free stream energy converter

Another machine in the HYLOW project is the free stream energy converter. It is a floating body with a v-shaped constriction at the bow. It has a funnel-shaped inlet leading to a narrow channel in the mid-section leading up to the rotor, and a diffusor at the outlet. The machine is shown in Figure 3-22 (HYLOW, 2012). The initial model tests were conducted by the project partner at the University of Southampton.

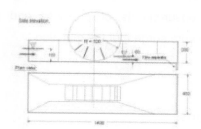

Figure 3-22: Free stream energy converter

A vertical blade is attached to the floating body at the outlet. This forms a region of low pressure and enables flow separation at the outlet. The units can be deployed in rivers and streams. The floating machine does not break the continuity of the stream or river.

3.2.4 Micro-turbines in water pipeline networks

The third machine type considered by the HYLOW consortium is the development of a micro turbine for power generation in water supply networks. For piston-type energy converters operating at pressures in the range of 10 kPa, losses are minimised and cavitation is avoided as the flow does not require acceleration. The configuration of the turbine was developed through model tests. Based on the results of the tests and with the help of prediction models, a large-scale model was constructed together with the necessary instrumentation for recording data. This prototype was then deployed for one year and recorded data was analysed.

With this turbine, electricity can be generated with a non-conventional technology. The volumetric positive displacement turbine design is based on high pressure turbines operated at pressures below 100 kPa. Using scale models, configurations for the turbine were tested for a maximum discharge of 12 l/s and a head difference of 1 m. A series of tests at various pressure differences and flow rates provided performance curves for the turbines and comparisons with the predicted performance. Based on the curves, a full scale prototype was constructed and installed in a water distribution network. The unit was monitored for one year with the instrumentation recording data like power supply etc. for analysis. The theory for a low-pressure energy converter using a piston was developed and the machine constructed. Measurements of pressure and power were recorded to investigate the behaviour (pressure surges) of the machine.

3.3 Design considerations

3.3.1 The power take-off mechanism

The layout of a power generation unit for HPM is shown in Figure 3-23. The figure shows the sequence of events from water entering the system to electricity exiting the system. The flow moves the blades, causing them to rotate. The blades are attached radially to the wheel shaft which is mounted on a metal framework in the channel or on a masonry or concrete structure. The flow exits the system by means of an outlet structure e.g. mill race.

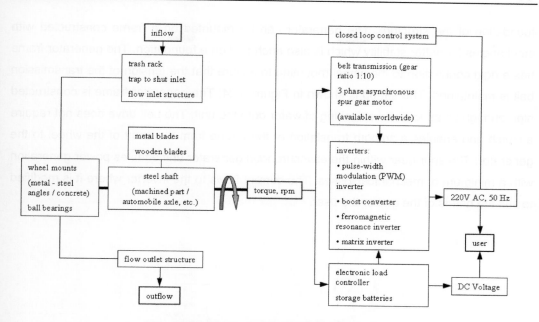

Figure 3-23: Design layout for power generation

3.3.2 Schematic set-up of power generation

The shaft of the machine is connected to a generator by a belt transmission, keeping a certain gear ratio. The generator can be a 3 phase asynchronous spur gear motor – a machine available worldwide. Alternatively, the shaft can drive a simple direct current (DC) generator. The generated electricity is then converted to an alternating current (AC) with a grid frequency of e.g. 50 Hz and supplied to the user (either locally or via the grid). In a simplified version, the DC generator can charge batteries and/or supply current directly to the user. In the case of power generation for supply at grid voltage and frequency, the machine needs to be fitted with a particular gear ratio to enable conversion to the grid frequency. This can be adjusted to suit the generator by running the transmission belt on a rim with the necessary diameter required for the gearing ratio. The wheel itself may be constructed with an automobile or truck wheel rim, the blades being welded perpendicular to the circumference. In the case of low wattage generation, the wheel can be constructed with alternative materials like steel motorcycle rims with wooden blades for a DC dynamo power take-off. The shaft of the water wheel is mounted on struts which are anchored in the foundation of the structure. The generator can be mounted on a frame constructed with steel angles for better stability which is also anchored on a foundation. The generator frame has a rigid connection to the wheel mounting to ensure that the tension of the transmission belt is maintained. The set-up is shown in Figure 3-24. The generator frame is constructed high enough so as to avoid splashing of water onto the unit. The belt drive does not require a clutch and enables a smooth translation of the torque from the shaft of the wheel to the generator. The spur gear within the asynchronous generator unit enables power generation with a minimum of mechanical losses. The current flows to the inverter where it is modified to suit the needs of the grid or the user.

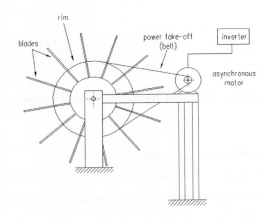

Figure 3-24: Schematic set-up of machine with generator unit and inverter

This simplified schematic set up points out the minimal requirements for a small hydropower runner-generator unit. The installation may require some concrete mounts depending on whether it is to run in a river or irrigation channel etc. Some additional parts like trash racks need to be installed before the intake into the wheel of chambers. Depending on the location and the availability of skilled labour at the location, parts of the machine may be manufactured locally. The required materials may also be available at the particular location.

Growing power demand is leading to the exploration of more small scale hydropower technologies with free surfaces for local power generation. Today, it is possible to design all the hydropower machines mentioned in this Chapter using Computer Aided Design (CAD) software tools. They are standard in commercial turbine design and are widespread in the manufacturing industry. The tools enable fast alterations in design based on analyses of e.g. the structural mechanics of components of the individual parts of the power generation unit. The fluid dynamics analysis for the design is enabled using CFD which can represent the free surfaces sufficiently. CFD enables an analysis of the flow situation and provides an estimation for power output, thereby enabling early corrections in machine design.

The absence of theoretical foundations for HPM machines is the impetus for research in novel power generation systems. In the following chapters, a theoretical foundation is developed for HPM machines based on CFD calculations.

Chapter 4
Computational Fluid Dynamics

4.1 Governing equations

4.1.1 Conservation principles

In the field of fluid mechanics the governing equations are based on conservation principles: mass, momentum-energy. These give rise to four conservation equations: conservation of mass and conservation of momentum/energy in x-, y-, z- direction (Batchelor, 1967). These equations are essentially derived from Newton's Second Law as first derived by Navier-Stokes in 1845. Generic exact analytical solutions for the N-S equations are not available. Exact mathematical solutions are only available for very simple flows. For practical engineering applications the N-S equations are discretised and solved by numerical iterative approximation methods.

4.1.2 Reynolds Averaged Navier Stokes equations (RANS)

The equations of relevance in this thesis are the RANS equations obtained by splitting the instantaneous variables (velocity u, pressure p) into two components: a mean and a fluctuating part (4-1, 4-2) (Reynolds, O. 1883 in (Rott, 1990)).

$$u(t) = \overline{u} + u'(t)\tag{4-1}$$

$$p(t) = \overline{p} + p'(t)\tag{4-2}$$

The Reynolds-averaged-Navier-Stokes equations (RANS) then become:

$$\frac{\partial U}{\partial t} + U \cdot \frac{\partial U}{\partial x} + V \cdot \frac{\partial U}{\partial y} + W \cdot \frac{\partial U}{\partial z} = -\frac{1}{\rho} \cdot \frac{\partial p}{\partial x} + v \cdot \nabla^2 \cdot U - \frac{\partial \overline{u'u'}}{\partial x} - \frac{\partial \overline{u'v'}}{\partial y} - \frac{\partial \overline{u'w'}}{\partial z}\tag{4-3}$$

$$\frac{\partial V}{\partial t} + U \cdot \frac{\partial V}{\partial x} + V \cdot \frac{\partial V}{\partial y} + W \cdot \frac{\partial V}{\partial z} = -\frac{1}{\rho} \cdot \frac{\partial p}{\partial x} + v \cdot \nabla^2 \cdot V - \frac{\partial \overline{u'u'}}{\partial x} - \frac{\partial \overline{u'v'}}{\partial y} - \frac{\partial \overline{u'w'}}{\partial z}\tag{4-4}$$

$$\frac{\partial W}{\partial t} + U \cdot \frac{\partial W}{\partial x} + V \cdot \frac{\partial W}{\partial y} + W \cdot \frac{\partial W}{\partial z} = -\frac{1}{\rho} \cdot \frac{\partial p}{\partial x} + v \cdot \nabla^2 \cdot W - \frac{\partial \overline{u'u'}}{\partial x} - \frac{\partial \overline{u'v'}}{\partial y} - \frac{\partial \overline{u'w'}}{\partial z}\tag{4-5}$$

The Reynolds stress terms on the right-hand-side $\left(\overline{u'_i u'_j}\right)$ require additional closure equations. This leads to the closure problem where the number of unknowns exceeds the number of equations. Approximation models (turbulence-closure models) are used to complement the equations. Every flow regime has a suitable turbulence model with different coefficients.

Numerical solutions can be classified in three main categories with increasing complexity. Turbulent flow at high Reynolds numbers show velocities which spread over several orders of magnitude. The resolution of the model enables the solution of the various scales. The categories are:

- Reynolds Averaged Navier Stokes equations (RANS)
- Large Eddy Simulations (LES)
- Direct Numerical Simulations (DNS)

4.1.3 Turbulence closure

"*Turbulence is not a well–defined problem awaiting a solution, but is a state of motion with innumerable different facets which depend on the context in which it occurs. We do not regard the state of laminar flow as a single problem; much less should turbulence be thought of as a single and soluble problem. The properties of turbulence are flow dependent, and therein lies the difficulty. Each turbulent flow field which is studied reveals new aspects of turbulence, and we are a very long way from being able to assemble a comprehensive physical description of this many–sided state of motion.*" G.K. Batchelor in (Froehlich, 2006).

Flow can be described in different manners as the state of the flow can range from laminar to turbulent. Laminar flow is stable and orderly and can be described mathematically. The layers in the flow are stable and run parallel. Laminar flows can be two or three dimensional and the flow occurs below a defined critical Reynolds number. Turbulent flow occurs when the inertial forces are much larger than the viscous forces in the flow. Turbulent flow is stochastic and three dimensional. Intensive mixing and diffusion occurs. Within the flow there is high vorticity and the flow is highly dissipative (the kinetic energy is transformed into heat-loss through viscosity at the vortices). It is highly irregular with the exception of flows with coherent structures.

4.1.4 Discretisation methods for numerical simulation

Computational Fluid Dynamics has made significant progress for solving the N-S partial differential equations. Various discretisation methods and numerical schemes have been developed. The most commonly used schemes in the field of CFD are (i) the Finite Difference Method, (ii) Finite Volume Method (FVM) and (iii) the Finite Element Method.

Developments in the field of solid mechanics of rigid bodies led to the finite element method (Zienkiewicz, 1971). The simplest method of discretising a conservation law is based on its differential or divergent form of the integral equations. The concept of the finite difference method method is finding a discrete approximation for the occurring derivatives and replacing the analytical derivatives with the discrete ones. This enables a numerical solution of a discrete problem as described by Hirsch in "Numerical Computation of Internal and External Flows, Vol. 1." (Hirsch, 1990).

Finite Volume Method (FVM)

The FVM approach was used in this thesis by performing numerical simulations using the ANSYS Fluent® software. The integral formulation of the conservation laws is the basis of FVM schemes. Fluxes through the discretisation element boundaries are evaluated instead of estimating derivatives. Different approaches exist for the choice of these fluxes. For convection dominated processes, one approach is the upwind method. Centred methods symmetric about the point where the solution is being updated are appropriate for systems in which information is travelling in all directions (Lecheler, 2009). The differential equations describing the fluxes here are elliptic equations. Fluxes in which information propagated in a certain direction are described by hyperbolic equations (supercritical flow in rivers). Parabolic equations, lying between the extremes (elliptic and hyperbolic) contain an advective term (representing the hyperbolic character) as well as a diffusion term (for the elliptic part). The relation between the advective and the diffusive part is described by the Peclet number (4-6) where v is the flow velocity, L a characteristic length and D the diffusion coefficient.

$$Pe = \frac{v \cdot L}{D}$$

4-6

For very small Peclet numbers, the diffusive part dominates and central discretisation schemes are used. For high Peclet numbers the advective term dominates and upwind methods are used. Upwind schemes use backward differences to discretise the advective term. The use of an explicit time discretisation and a backward difference scheme in space leads to the first-order upwind method for the advection equation (4-7).

$$\frac{1}{\Delta t}\left(u_i^{n+1} - u_i^n\right) + \frac{v}{\Delta x}\left(u_i^n - u_{i-1}^n\right) = 0$$

4-7

Figure 4-1 shows the discretisation with the backward difference scheme in which the dashed lines represent the time and space levels and the boxes show the control volumes.

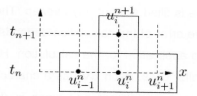

Figure 4-1: Upwind discretisation

The first-order upwind method for the advection equation (4-8) is conditionally stable.

$$u_i^{n+1} = u_i^n - \frac{v \cdot \Delta t}{\Delta x}\left(u_i^n - u_{i-1}^n\right)$$

4-8

The dimensionless Courant number given by the Courant-Friedrichs-Lewy condition (4-9) describes the ratio of the physical velocity v over the grid velocity $\frac{\Delta x}{\Delta t}$. For $Cr = 1$ the solution is exact, for smaller Courant numbers the solution is stable but a large amount of numerical diffusion is introduced.

$$Cr = \frac{v \cdot \Delta t}{\Delta x}$$

4-9

FVM methods enable a straightforward derivation of the cell-centred FVM equations. The method is easy to implement and mass-balance is met for every cell and therefore for entire domain. The method is not restricted to structured grids. Due to pressure gradients parallel to the interfaces, the evaluation of fluxes may be complicated on unstructured grids. FVM methods consider volume-averages. The evaluation of fluxes due to pressure gradients parallel to the interfaces may lead to non-symmetric system of equations.

4.1.5 Discretization in space and time

In CFD three basic types of grids can be generated depending on the application: structured grids, unstructured grids and a combination of both, hybrid grids. The control volume for the fluid is defined over the geometry of the model. This volume is discretized into smaller units consisting of cells or control volumes (CV) and nodes in such a manner so that the entire fluid volume is filled without any overlap. The calculation within the flow occurs at these points and the error of the calculation depends on the quality of the mesh. Mesh density depends on the application for the calculation. High velocities require a high resolution of the grid i.e. smaller CV's. Each CV has a characteristic point e.g. cell centre (Lecheler, 2009). The difference in different grid types is depends on the relationship between the CV's. Variations in geometry leading to alterations in the prevalent physical conditions are refined to allow for the local solution. These zones are parts of the whole fluid volume and can be refined, adapted or coarsened as required. Grids may be structured or unstructured or a combination of both. Within a structured grid the cells of a topology are similar. The CV' are regular and are ordered in a raster in square or rectangular shapes in 2D space. In 3D space the CV's consist of hexahedrons. In the case of complex geometries the CV's can be adapted individually to best fit the topology. Although structured grids have the big advantage of structured calculation schemes and proper resolution of corner flow boundary layers, the disadvantage of high generation efforts and point numbers – unless methods like hanging nodes or overset grids are used – limit their applicability for complex geometries (Gansel *et al.*, 2014). In unstructured grids the relationship between CV's are variable. In the 2D case, unstructured grids usually consist of triangular elements. The mesh can be refined by defining the number if points on a curve. The triangles can also be generated together with quadrilateral elements thereby enabling a versatile structure. In the 3D case, the grid is composed of tetrahedra, hexahedrons or prisms. All three basic elements can be combined. Complex geometries can be discretised with unstructured mesh. In the meshing process points of an unstructured surface are extruded from the walls (Gansel *et al.*, 2014). Hybrid meshes are a combination of structured and unstructured grids. Two or more independently meshed zones in the geometry are merged into one mesh with the tetrahedral or prismatic parts connected directly to each other.

4.2 Free surfaces, volume of fluid (VoF) approach

4.2.1 Multiphysics approach to free surface flows

In hydrodynamics several models derived from the Navier-Stokes equations for incompressible free-surface flows exist. 1D, 2D and 3D models can be categorised as hydrostatic and non-hydrostatic types. In descending order of complexity, for the 3D case either the free surface Navier-Stokes or the hydrostatic 3D shallow water equations may be considered; in the 2D situation the Boussinesq, Serre or Saint-Venant equations can be adopted. The latter set of equations may be considered in 1D cases. A full 3D model of the problem at hand can capture all the physical features. Thus problems relating to a wide spectrum of space and time scales related to the presence of different physical phenomena can be covered.

The mathematical model can consist of two fluids (two continua) with a sharp interface where the free surface is represented by a boundary in a moving mesh or as a jump in properties in a single continuum (step-change for density, viscosity). The free surface is moving with no mass flux through the surface. It is however momentum compatible. The surface capturing model is a single continuum approach where the interface is not discrete. VoF uses this approach. Advantages are a natural handling of the interface breakup and a static mesh which makes the simulations more efficient. The theory producing waves in water by disturbing the surface has been investigated in a paper by Cauchy (Cauchy, 1827). It is commonly referred to as the Cauchy wave problem. This problem is of particular importance in the field of fluid dynamics as it can be implemented in modelling surface waves and their propagation (Sommerfeld, 1992). Substantial mathematical effort and refinement is required to obtain results. Research has been done on the numerical simulation of flows with free surfaces or moving boundaries, the majority of which have dealt with slosh motion in closed containers (Ibrahim, 2005), wave impact problems (Kleefsman and Veldman, 2004) or wave resistance of ships (Sung and Grilli, 2005).

An overview of various different methods for free surface flows and moving boundaries is given in (Lakehal, Meier and Fulgosi, 2002). Slosh and wave problems as well as the tracking of the interface of two immiscible fluids like air and water have been widely simulated using the VoF approach. This approach became popular due to the SOLA-VoF code presented in a paper by Hirt and Nichols (Hirt and Nichols, 1981). A liquid volume fraction field, usually named F, is used to describe the fluid configuration (phase information). All phase information in VoF methods is stored in the F-field. Interface locations are approximated from a so-called interface reconstruction. A liquid indicator function F is used in VoF methods, with F = 1 for liquid and F = 0 for void (or gas) regions. This F-Field is an additional advection equation as well as a property of the fluid with which it moves:

$$\frac{\partial F}{\partial t} + (u \cdot \nabla)F = 0$$

4-10

Local density is linearly related to the liquid function indicator F by the following equation:

$$\rho = \rho_V + F(\rho_1 - \rho_V)$$

4-11

with the material parameters void-density ρ_V and liquid-density ρ_1. In the case of water and air, the latter is treated as void region with uniform pressure.

4.2.2 Surface tracking

The ANSYS Fluent® VoF model is a surface tracking model is a two fluid approach with a separate mesh for each phase. Mesh motion is defined as a boundary condition with the mesh adjusted for the motion of the free surface. Surface tension is defined by a pressure jump on the free surface. This method enables a precise modelling of the free surface with a sharp interface and mesh compatibility through boundary conditions. The interface is defined by a regular set of marker points located on the interface. The position of the interface is approximated by a piecewise polynomial interpolant (Hyman, 1984). Hence the domain is divided into connected regions by the time-dependant interfaces. To account for discontinuities, the solution along the interface and at marker points may take on multiple values. The marker points themselves can be represented by a height function from a defined reference surface or by a parametric interpolant. The former is easily implemented but is restricted if the surface becomes multivalued with respect to the reference surface. Sub-grid structures in the interface can be captured in two and three dimensions with both methods in high resolution.

4.2.3 Volume tracking methods

Interactions of many different smoothly varying interfaces can be estimated accurately using volume tracking methods. The interface tracking equations have the same dimension as the underlying PDEs but do show not the same sub-grid resolution as surface tracking methods. The interface data is stored and the equations solved in the cells along or near the interface resulting in an increase in computational complexity but an improvement in efficiency when the interfaces are well separated (Hyman, 1984). The marker and cell volume tracking method consists of marker particles scattered to identify the material regions. The materials and the particles are transported using a Lagrangian approach. When found within a cell, the marked material is indicated to be present in the cell. The boundary is calculated by using the densities of the marker particles in the mixed cells and marker particles of the other materials. Interfaces are reconstructed using the densities of the particles in surrounding cells. Efficiency is improved by a denser scattering of initial marker particles. For improved performance, more particles than computational cells are needed thereby increasing computational costs. These drawbacks are eliminated by introducing fractional marker volume methods called the volume of fluid method. Here the fractional volume of each material within a cell is calculated. The values of the fractional volume range from one (completely filled with material) to zero (no material). Cells that contain fractional volumes define the interface (Figure 4-2) (Hyman, 1984). The interface position is reconstructed for every time step using the fractional volume of the cell and the surrounding cells.

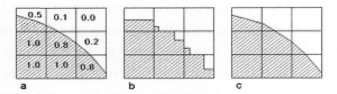

Figure 4-2: Interface construction between two regions: (a) actual situation (b) rectangular approximation (c) piecewise-linear fractional volume method

The reconstruction of the interface with the volume fraction can be represented by unions of rectangles or piecewise polynomials within the cells. Finite-difference approximates based on the neighbouring fractional volumes can be used to estimate the curvature of the interface. More complex methods to approximate the interface are sensitive to numerical diffusion and can be improved by imposing monotonic and convexity constraints. Various reconstruction algorithms are used depending on the application and the demands of sub-grid scale structure. The reconstruction algorithm can be supported by maintaining and evolving a point on the interface in each fractional volume cell. The marker points are moved with the interface. The same methods described for surface tracking can be used to advance the interface once it has been constructed.

For machine with symmetrical geometries, simplified 2D models can be used to analyse the flow in one plane and can provide estimates for e.g. mesh parameter. and solver settings. With increasing hydropower machine complexity, geometries should be modelled and run in 3D. The next two chapters will explore details of 2D and 3D calculations.

Chapter 5
Application of 2D CFD modelling

The modelling of the HPM machine was done in two steps. Since flow conditions within a channel are complex in 3-D, the initial step consisted of setting up a 2D model for the wheel with straight blades. Here the processes involved in the filling of the chambers were visualised. In a further step, a 3D turbine and channel was modelled to enable the implementation of asymmetrical elements in the machine geometry (e.g. curved blades). In consideration of the free water surface and asymmetric machine geometry, the whole machine within the channel was modelled.

In order to model the machine setup, the commercial code ANSYS FLUENT® was selected along with the mesh generator ANSYS ICEMCFD®. Both products are marketed by ANSYS Inc®. This software was chosen for the following reasons:

- hexahedral and tetrahedral meshing
- capability of modelling quasi two and three dimensional flow
- steady-state or transient flow analysis with rotating mesh
- capability of modelling multiphase flow with free surfaces using the Volume-of-Fluid (VoF) method
- choice of turbulence models

The aim of the numerical models was to understand better the physical processes involving free surfaces in the hydrostatic pressure machine using the CFD simulation software. This included implementing direct simulations using VoF methods to model the free surfaces. Models using VoF methods provide for a realistic environment and are more complex than single-phase models (where the flow is simulated in a "closed" chamber). Multiphase models together with mixture and Eulerian models enable a visualisation of flow in machines and can help in explaining physical phenomena which occur in these processes.

The numerical model of the turbine was generated using the ANSYS software programs ICEMCFD and FLUENT 12.0.1. These tools enable mesh generation and CFD for the analyses. The generated meshes can be created in various formats with varying complexity such as structured, multiblock, unstructured, hexahedral or tetrahedral meshes as required. The program also enables a geometry import from various CAD formats. Points, curves, surfaces can be modified or created to suit the necessary mesh generation. The created mesh can once again be saved under different formats and can be translated to formatted input files for particular solvers. The 2 - equation $k - \varepsilon$ turbulence model was used with standard wall functions. Transient runs were made which allowed for the use of the geometric reconstruction algorithm with surface tracking. For pressure discretization, the body force formulation was used as gravitational forces were present. Momentum and turbulent kinetic energy were handled with first order schemes. As the runs were transient, the pressure-velocity coupling was realised with the PISO algorithm. The initial 2D runs were made to test the behaviour of the numerical model with the free surface. This model was set up to understand the processes occurring in one plane within the channel. The variations in torque at the shaft in the experiments could be explained with the visualisation of the filling and emptying of the chambers during operation. The model was verified with results from an experimental setup. In a further step the boundary conditions were varied and the calculated torque was compared with experimental results.

5.1 2D Setup: Preprocessing

5.1.1 Domain

The original model that was constructed is shown in Figure 5-1. This model consists of a rectangular 2D computational domain with an inlet at the left, a bottom wall, an outlet on the right. The top consists of a symmetrical boundary. The wheel with 12 blades is centrally located (purple region). The domain consists of two parts - the wheel with blades and the rest of the domain thereby allowing for rotation of the wheel.

Figure 5-1: Computational domain

Successful runs were made with the initial model and results showed that the main effect on the position of the free surface in and around the turbine was the entry of the blades into the water. This would indicate that the distance from the upstream boundary to the turbine is not as critical as in other applications. As long as the depth is set correctly to match experiment, this distance can be reduced which brings a saving in computing time. A new domain was therefore adopted that reduces the length.

5.1.2 Mesh

The grid, with the reduction in domain size, was relatively coarse in all areas. However, simulations showed that the grid resolution was sufficient to ensure convergence with Volume of Fluid (VoF) schemes and the mesh density was sufficient to ascertain the main features of the flow in the machine and could be used to develop the model further. An option in terms of grid refinement is to refine the grid dynamically around the position of the free surface. However, this will incur a heavy computational load and the improvements in free surface resolution may not lead to significant changes in the moments calculated. This option was therefore not pursued. The mesh used for the simulations of the 2D turbine consists of 40370 elements of the types *lines* and *quadrilaterals* with a total of 38235 nodes. Two separate meshes regions were defined generated, which allows for the definition of the rotating and the stationary zones of the grid.

To match the experimental set up of the test rig, the grid was modified directly below the wheel. The grid is shown in Figure 5-2. The mesh consists of two main regions: the central region of the wheel with hub and blades and the surrounding region of the computational domain. The mesh was coarsened with increasing distance from the wheel. The region of the wheel rotates at the defined operational speed.

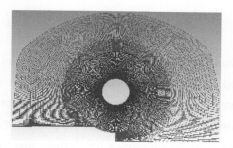

Figure 5-2: Computational grid of the 2D model

To ensure that the Courant-Friedrichs-Lewy criterion is fulfilled, regions where high velocities were expected have a refined mesh. These regions were located at the interface separating the two regions (Figure 5-3). On the computational domain side of the interface, the adjoining cells were gradually coarsened, thus saving computational time.

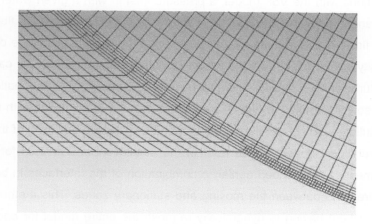

Figure 5-3: Refinement at mesh interface and gap

The interface between the domain and the rotating region comprised of two zones: the interface zone on the wheel side of the mesh and the interface on the domain side of the mesh. Both these zones were merged into a common interface. The rotating zone in the domain contained the hub and blades of the machine. The blades were represented by lines having two "sides": the blade and blade-shadow.

5.1.3 General settings

As in the case of VoF calculations, a pressure based solver was used in the simulations. This solver is implemented for low speed incompressible flows with velocity fields being calculated from the momentum equations. For the VoF parameters, the explicit model with open channel flow was taken. For the two phases in the domain, the materials air and water were chosen with standard material properties for density and viscosity. Turbulence was described using the standard 2-equation k-ε model with standard near-wall functions. The cell zones for the rotating part of the mesh and the stationary domain were defined. The 2D model of the wheel was contained in a zone with the blades and hub modelled as lines. The bed of the channel was modelled as a wall with a constant roughness.

Boundary conditions

For the simulations with the VoF model, a pressure inlet in with a given water height (free surface level) and bed level of the channel were defined, thereby giving the flow velocity at the input. The flow velocity was calculated based on experimental discharge data supplied by the University of Darmstadt from the flume. Turbulence at the inlet for the cases run was specified by turbulence intensity and the hydraulic diameter. For the outlet boundary condition, the solver permits the use of a pressure outlet in combination with the pressure inlet. For the outlet, a gauge pressure, free surface level and the bed level of the flume was defined. The rotational speed for the rotor in the simulation was defined for the rotating region of the mesh. As described earlier, a combination of the interfaces in both zones of the mesh differentiate between the moving and stationary zones. This method employs a moving mesh without having to dynamically re-mesh the entire model for each time step. Rotational speed and discharge for the models was based on experimental data.

Initial conditions

The initial conditions for the pre-processor parameters were calculated based on the data like velocity of the flow specified at the input. To define which parts of the domain contain water, all zones were first initialised with air. Subsequently, the regions of the mesh to be filled with water were selected according to the specified upstream and downstream water levels, and were defined as water by giving a VoF value equal to 1 (volume fraction = 1).

Time step

As mentioned in Chapter 4, the time step is calculated using the Courant number which sets a limit on the step size based on the cell size in the mesh. Initial runs were performed with different time step size Δt from 8.10^{-5} s to 1.10^{-3} s. The effect of time step size was seen in the change in run time but the difference in the results was not significant. Similar initial simulations were performed with different mesh refinement. Here again an increase in mesh density was reflected in the run time with no significant changes in the results for torque.

5.1.4 2D Cases

Based on the experimental data obtained, four cases with head differences Δh varying from 0.215m to 0.619m were chosen. As the origin of the coordinate system lies in the centre of the wheel, the upstream water levels may have negative values. Values for the upstream and downstream were values relative to the axis of the runner. Table 5-1 shows the simulation cases with upstream and downstream water levels along with the discharge. The water levels were measured from the axis of the hub. The geometry and mesh were run with the data supplied for upstream and downstream depth, discharge (set by a velocity at the upstream boundary) and the rotational speed of the wheel, thereby fulfilling boundary conditions requirements for the model.

In Case 1, two simulations were run with a discharge of 60 l/s. Two runs were performed in Case 2 with a discharge of 100 l/s. Case 3 and Case 4 had discharges of 130 l/s and 180 l/s respectively. The different cases were analysed in terms of water depths, discharge and torque in order to monitor the processes occurring within the flow.

Table 5-1: Initial conditions for 2D simulation models

	Discharge [m³/s]	Upstream level h_{up} [m]	Downstream level h_{down} [m]	$\Delta h = h_{up} - h_{down}$ [m]	rpm [1/min]	
	0.060	0.010	-0.420	0.430	2.47	
Case 1						
	0.060	-0.002	-0.213	0.211	3.11	
	0.100	0.001	-0.411	0.412	4.89	
Case 2						
	0.100	0.004	-0.504	0.508	5.23	
Case 3	0.130	0.204	-0.212	0.416	6.5	
Case 4	0.180	0.203	-0.416	0.619	8.77	

5.1.5 Postprocessing

Reference values

Torque is calculated from the force components acting on the faces of the hub and the blades of the rotating wheel. the total force comprising of the pressure force component and the viscous force component (Ansys Fluent, 2012). The forces are calculated along a vector with the corresponding pressure and viscous force components in the specified direction for the faces. For each of the wall zones, the net of the values for forces (pressure and viscous) and the coefficients (C_F) are calculated. The expression for calculating the dynamic force F for an area A is given in the following equation

$$F = \frac{\rho \cdot v^2 \cdot A}{2} \cdot C_F \; [N]$$

5-1

Using the calculated force at each face, the software ANSYS FLUENT®, computes the moment for each face (Ansys Fluent, 2012).

$$M_T = F \cdot r = \frac{\rho \cdot v^2 \cdot A}{2} \cdot C_F \cdot r \; [Nm]$$

5-2

The power generated can be calculated as the product of torque M_T and the rotational speed ω as given in equation 5-3. This is the mechanical power P_{mech} available at the shaft of the machine.

$$P_{mech} = M_T \cdot \omega \; [W]$$

5-3

As shown earlier, for given discharge Q, and head difference $\Delta h = (h_{upstr} - h_{downstr})$, the theoretical power P_{theor} is given by

$$P_{theor} = \rho \cdot g \cdot Q \cdot \Delta h \; [W]$$
5-4

The efficiency η is given by the ratio of the power available at the shaft of the wheel P_{mech} to the theoretical hydraulic power P_{theor} available in the system.

$$\eta = \frac{P_{mech}}{P_{theor}} = \frac{M_T \cdot \omega}{\rho \cdot g \cdot Q \cdot \Delta h}$$
5-5

5.2 2D Analysis

The use of an implicit VoF method reduces computational time. However, when using this, the solution in Figure 5-4 (a) was obtained. From animations of this simulation it could be seen to proceed as would be expected, but later in the simulation water could be seen to rise in the separate chambers as shown in the image. This appears to be due to poor resolution of the free surface which is spread over many cells. The calculated torque also reflects this inaccuracy. In view of this the VoF update method was switched to the more accurate explicit mode and Figure 5-4(b) shows that this produces a much better resolution of the free surface and consequently a more realistic solution. The need to use the explicit VoF update imposes a computational cost.

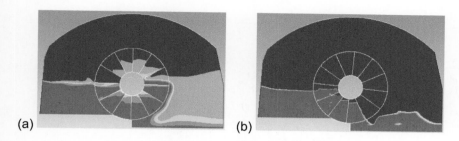

(a)

(b)

Figure 5-4: Contours of density a) implicit, b) explicit

5.2.1 Case 1

Results of the first case (discharge 60 l/s) made with head difference of 0.430 m are shown in Figure 5-5. The torque generated shows a series of peaks and troughs with smaller fluctuations between the major peaks.

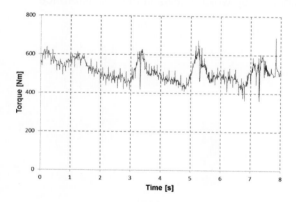

Figure 5-5: Torque on rotor Case 1, Δh = 0.430m

For the same Case 1 with a smaller head difference, the peaks and troughs appear similar to those of the previous run, the difference being the absence of most of the minor fluctuations. This indicates a smoother running of the wheel at a lower Δh.

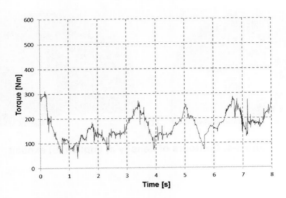

Figure 5-6: Torque on rotor for Case 1, Δh = 0.211m

In Case 1 the mean torque calculated shows lower values than the measurements as shown in Figure 5-7. The case with the smaller Δh shows about 25% lower values as compared to the second setup which shows about 7% difference.

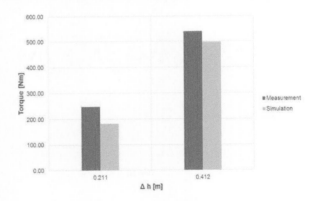

Figure 5-7: Comparison between measurement and simulation

5.2.2 Case 2

Results of the Case 2 simulations with an increase in discharge (100 l/s) showed similar characteristics in the recorded torque. The moment on the wheel as calculated in the simulation for a head difference Δh = 0.412 m is shown in Figure 5-8. This indicates that whilst the moment is not steady, the variation has a relative stationarity. The repetition of the series of peaks as seen in the figure indicate that the fluctuation in torque is a result of a series of processes. This behaviour was seen in all simulation results.

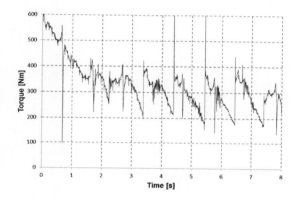

Figure 5-8: Simulation torque for Case 2, Δh = 0.412 m

Figure 5-9 shows a further example of the torque generated in the simulation showing periodic characteristics. The amplitude and frequency for torque depends on the head difference and the rotational speed for the given operating case.

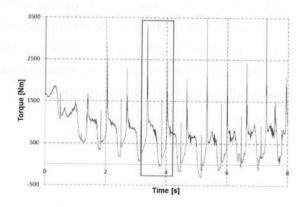

Figure 5-9: Fluctuations in torque generation

To visualise the flow conditions at the wheel for the time between two peaks in generated torque, the region marked in red in the above Figure 5-9 was chosen. The progression for torque in this region is shown in the following Figure 5-10 shows torque one second of simulation time.

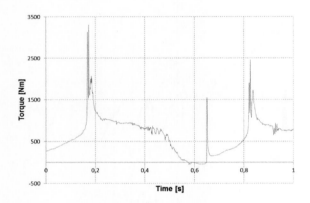

Figure 5-10: Simulation time = 1 s

The above diagram shows two significant peaks at t ≈ 0.17s and t ≈ 0.83s in the curve for the torque generated in the simulation. For a given rotational speed, the peaks in the torque indicate that two processes occur during one cycle i.e. in the time between two following blades enter the upstream water surface. The following series of images were captured during the course of the above figure every 0.05 seconds in the simulation. Figure 5-11, a) shows water entering the chamber of the runner at the beginning of the curve for torque (above Figure 5-10). Water flows into the chamber above the submerged blade towards the hub, increasing the torque at the shaft. On striking the hub, the water is reflected. The change in direction at the almost completely filled chamber produces a peak in the torque at the shaft. This torque drops rapidly as the filled chamber submerges.

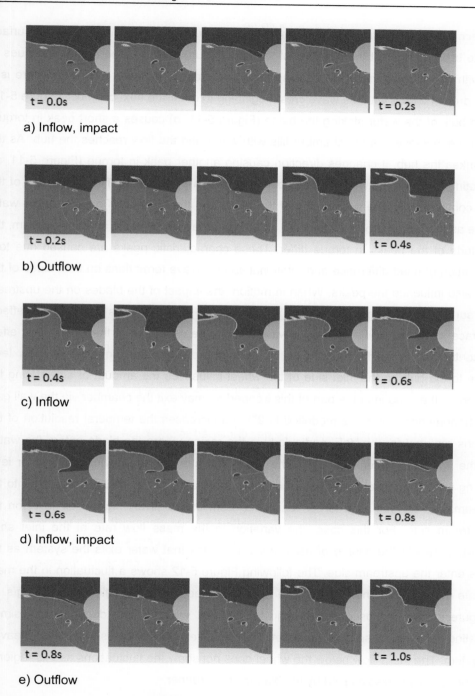

a) Inflow, impact

b) Outflow

c) Inflow

d) Inflow, impact

e) Outflow

Figure 5-11: Image series for 1 s simulation time

The second blade entering the water at the upstream side of the wheel causes a surface wave to build up against the direction of the flow (Figure 5-11, b). The torque continues to drop with the submerging blade causing the wave crest to increase in height. There is a further drop in torque with the wave crest beginning to enter the next chamber (Figure 5-11, c). The bulk of the water striking the blade (Figure 5-11, d) causes a short peak in torque. The torque increases as the chamber fills with water and the flow reaches the hub. As the flow strikes the hub, it changes direction causing another peak in torque (Figure 5-11, e), completing the cycle between the two peaks shown in Figure 5-10. The movement of the wheel continues with the submerging of the blade and the build-up of a wave at the water surface as the torque drops. Depending on the water levels upstream and downstream, the magnitude of the peaks in torque differ. These characteristic peaks are caused due to a combination of head difference and rotational speed. Wave formations on both sides of the wheel also influence the peaks. When in motion, the impact of the blades on the upstream water surface causes some amount of air to enter the chamber. This dampens the effects of surface waves as well as pressure waves within the chambers of the wheel. This effect also contributes to the fluctuations in torque on the shaft of the runner. The water level relative to the hub on the inlet side of the wheel influences the amount of air entering the machine. In the experiment, a part of this trapped air may exit the chamber via the wall gap. This phenomenon cannot be modelled in 2D and increases the temporal resolution of the solutions required for the VoF, particularly in the region where the blade enters the water. With the upstream water level up to the top of the hub and the downstream water level reaching to the bed level of the hub, the blades incur a loss in moment upon entry into the upstream side as well as a larger resistance from the relatively high water level on the downstream side. For this case, the variation in the mass flow rate at the inlet show negative values in the course of oscillations, indicates that water exits the system as the blades enter the upstream side. The following Figure 5-12 shows a fluctuation in the mass flow rate at the inlet of the domain. The value assigned to the simulation was 100 kg/s. As the figure shows, the oscillation around the defined value is time dependent. A similar fluctuating behaviour is seen at the outlet. The negative value indicates that mass is leaving the system. The mass flow below the wheel does not show the fluctuations as this region is not influenced by waves caused by the blades of the runner.

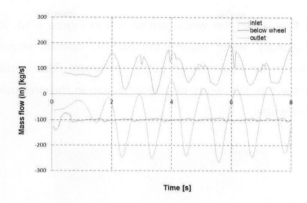

Figure 5-12: Mass flow in the system, Case 2, Δh = 0.412 m

Water levels

In the experiment the water depth in the flume was measured in the middle of the channel at a distance of 5.27 m upstream and 1.89 m downstream of the shaft of the runner. This region has not been modelled in the 2D simulations as the flume is wider in this region. The water levels at the inlet and outlet of a simulation model are shown in Figure 5-13. These levels also show periodic fluctuations around the defined value. The fluctuations are caused by the blades of the wheel.

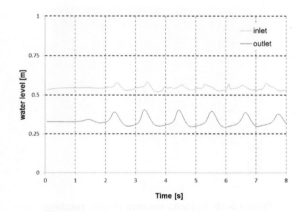

Figure 5-13: Water level at inlet, outlet for Case 2

Velocity vectors

The velocities in the flow within the wheel are shown in Figure 5-14. In the case of the wheel being kept stationary, water leaving the partially opened segment of the wheel recirculates at the step in the channel bed below the wheel. The velocity vectors in this case (Figure 5-14) show that this leads to a recirculation in the previous segment on the downstream side of the wheel.

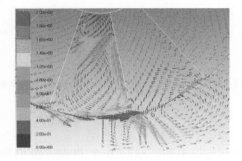

Figure 5-14: Velocity vectors in gap, stationary

As the wheel rotates, the chambers on the downstream side begin to empty, thereby increasing outflow velocities. Figure 5-15 shows velocity vectors for the runner with the chambers emptying downstream. The zone of recirculation below the chamber has shifted considerably further as a result of water leaving earlier segments still partially filled.

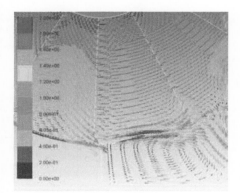

Figure 5-15: Velocity vectors in gap, rotating

The following figures show the torque on the rotor for case 2 with Δh = 0.508 m.

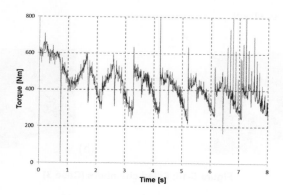

Figure 5-16: Torque in Case 2, Δh = 0.508 m

5.2.3 Case 3

Results of the simulation of Case 3 for a discharge of 130 l/s with Δh = 0.416 m are shown in Figure 5-17. The results are not realistic and show a negative torque on the machine. This is caused by water remaining in the chambers as they rise and causing a negative moment that outweighs the positive moment generated earlier in the rotation. It is felt that the late emptying of the chambers is caused by the 2D assumption.

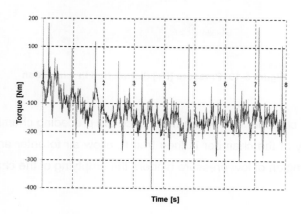

Figure 5-17: Torque on rotor for discharge 130 l/s

In this simulation, the contours of the phases show that water does not leave the chambers and is transported back to the upstream side of the model. Figure 5-18 shows the partially empty chambers of the wheel.

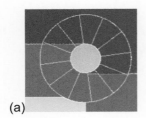
(a)

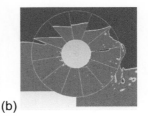
(b)

Figure 5-18: Filled chambers (Case 3)

In the 2D model, air cannot enter or exit the chamber in the transverse direction. All the air entering a chamber can only do so at the outer boundary of the chamber. Water leaving the chamber is replaced by air entering the chamber. Figure 5-19 shows the pressure distribution around the rotor. At regions of low pressure within the chambers (blue), the preceding chambers still contain water.

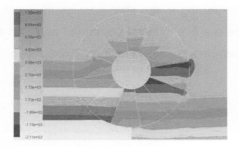

Figure 5-19: Pressure [Pa] distribution in simulation

As a result of this, a chamber can only run dry once the preceding chamber stops releasing water. The boundary of the chamber is then free to allow air to enter and water to exit the chamber. Here the region of low pressure hinders the emptying of the chamber.

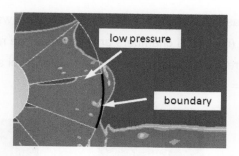

Figure 5-20: Chamber boundary with low pressure region

A change in the settings for the properties of air from constant density to ideal gas to allow for some compression and expansion at the downstream side of the wheel was tested. The effect of these settings show a drop in the water level on the upstream side with values for torque remaining negative.

5.2.4 Case 4

Simulations for Case 4 with a discharge of 180 litres/second ($\Delta h = 0.619$) also show a negative torque being generated as a result of the chambers not emptying (Figure 5-21).

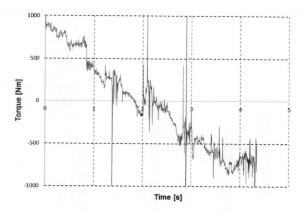

Figure 5-21: Torque for Case 4, discharge 180 l/s

Similar to the previous case, the following Figure 5-22 (a) shows the initial state of the simulation. In Figure 5-22 (b) the full chambers are shown for the run with air with constant density at t = 4.57 s. Figure 5-22(c) shows the drop in water level at the inlet with air being defined as an ideal gas (t = 1.57 s).

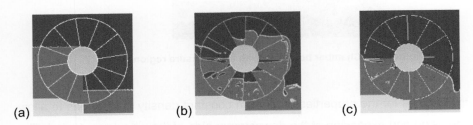

(a)   (b)   (c)

Figure 5-22: Water level drop at inlet (Case 4)

5.3 Comparison of results

For the given fall head in the different cases considered, a comparison of efficiencies of the machine from experimental data and simulations are shown in Figure 5-23. The simulations show an underestimation in the efficiencies of the cases run. This could be due to effects like transverse flow being hampered by the two-dimensionality of the model.

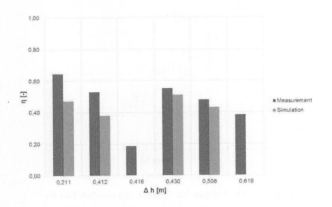

Figure 5-23: Comparison of efficiencies

The difference in torque and the conversion of this torque in terms of fall head are listed in the following Table 5-2. These indicate a difference in water level at the inlet of the computational domain in the order of a few centimetres. This difference in depth shown in the table indicates that the power loss lies in the range of a few watts and does not lead to a significant change in terms of power generated. These differences could be due to the disturbances caused by the change in channel-width as well as other effects like wave formation or splashing and the influence of the walls of the flume at the wheel.

Table 5-2: Differences in torque in terms of fall head

	Δ h [m]	Torque difference (experiment - simulation) [Nm]	Power difference [W]	Water level difference [m]
Case 1	0.211	66.79	21.75	0.013
	0.430	41.90	10.84	0.006
Case 2	0.412	119.74	61.32	0.060
	0.508	43.28	23.70	0.023

The 2D runs showed a quantitative estimation of power generation for the simulations with discharge of 60 l/s and 100 l/s. The higher discharges could not be modelled successfully making it necessary to model the machine in a 3D environment.

5.4 Discussion: 2D case

A plan view (not to scale) of the experimental set up of the flume is shown below (Figure 5-24). The 2D computational domain modelled represents a part of the whole system of the rotor in the flume, from the points of water level measurements upstream and downstream of the wheel. Factors that influence the 2D model are addressed hereafter.

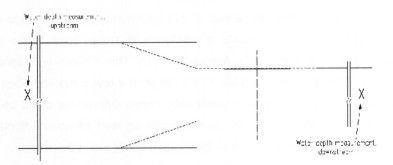

Figure 5-24: Plan view of experimental set up

Measurement of the water levels

The water level for the machine with straight blades was measured at points 5.89 m upstream and 1.29 m downstream from the position of the shaft of the machine. These regions were not modelled in the 2D model as the change in flume breadth cannot be represented in 2 dimensions. As the process is highly dynamic, local fluctuations in water levels at the wheel as well as flow velocities cannot be considered in the present model. These effects may well lead to changes in the values for torque and explain discrepancies. Other effects like wave reflection on the sides of the flume and at the wheel blades on entry into the upstream water surface cannot be accounted for in the 2D model.

Gap losses

In the experiment, the gaps between the rotating wheel and the walls and bed of the channel have been defined in millimeters. The 2D representation considers only the gap between the wheel and the bed of the flume. As a result of not being able to define the gap in the 2D model, water the amount of water exiting the chambers on the downstream side cannot be accounted for. The images of the simulations show that some amount of water is transported in partially filled chambers from the downstream to the upstream side of the rotor, thus accounting for a loss of torque. The influence of the wall gaps along the sides of the channel are represented in the 3D model. The 2D simulations however show that the entrapment of air in the chambers can be a hindrance in the 3D model with narrow lateral wall gaps.

Flow conditions at the inlet

Conditions imposed on the flow at the entry into the flume of rotor breadth indicate that there a considerable amount of energy is dissipated in the form of turbulence etc. (Figure 5-25). Through the narrowing of the flume, the effects of the jump in breadth can be carried up to the rotor. Surface wave formation in random directions is possible, as propagation is influenced by the impact of the blades on the upstream free surface. These waves move against the direction of flow and cause further losses in the form of head losses. The reduction of the cross section of the flume and its effects on the flow velocities directly upstream of the rotor cannot be adequately described in 2D as the change occurs in the z-direction. Hence, a change in the velocity profile immediately before the wheel can be expected. The oscillation of the values of torque on the shaft of the machines is due to the frequency of the wheel blades striking the water and partially due to water waves being reflected by the inlet and outlet.

Flow conditions at the chambers of the wheel

Further, the point of entry of the blades into the water on the upstream side, in reality, is a three dimensional issue as the wall gaps between rotor would influence the filling rate of the chambers, allowing for a quicker filling of the chambers and a faster loading of the blades. Water can gush into the chambers through the wall gaps, and as this cannot be represented in the 2D model, can lead to lower torque values in the simulations. On the downstream side of the machine, this process would allow for a quicker emptying of the chambers with air entering the chambers through the gaps.

Turbulent losses

Figure 5-25 shows a sketch of the flume with arrows indicating flow velocities around the entry into the computational domain. For the 2D simulations, the part of the flume with constant breadth upstream and downstream of the wheel was modelled. With the wheel in motion, deviations in water levels arise with wave propagation in different directions (upstream as well as downstream). Based on the geometry of the flume, all these factors cannot be included in 2D simulations. Hence the influence of these factors is not represented in the 2D model. Fluctuations in the water level immediately before the wheel are also not accounted for at the inlet of the model. Transverse waves may cause a local increase in head which may account for higher values for torque in the experiments.

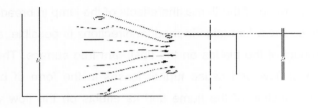

Figure 5-25: Losses at entry to the wheel

2D limitations

The 2D model in ANSYS FLUENT® calculates the flow for a unit width of 1 m of the machine. In the model, the blades of the wheel are represented by lines which are meshed from both sides. The software recognises each side (of the blades) as *blade* and *blade-shadow*. Torque is calculated for both the sides of the blade. A more accurate method to calculate torque would be to assign the blades a finite thickness, thereby defining the blades as a part (zone). This would enable the software to distinctly identify the zones of the blades and the corresponding values of the cells. Each side of the blade would have its unique cell bordering it.

The simulations show that the 2D water wheel simulations are valid for low flow situations. The emptying of the chambers of the wheel in reality takes place in three dimensions, with air entering the chambers as the blade rises off the free water surface as well as air entering the chamber through the wall gaps. Hence, the limitations of 2 dimensional modelling do not allow for these processes to be sufficiently resolved. The 2D models however provide an assessment of required mesh refinement, boundary conditions, solver settings and solution methods.

For machine with symmetrical geometries, simplified 2D models can be used to analyse the flow in one plane and can provide estimates for e.g. mesh parameter and solver settings. With increasing machine complexity and asymmetries, geometries should be modelled and run in 3D. With an increase in computational capacity with every new generation of computers, sensitive conventional turbines hydropower can be modelled to a high accuracy. Models can be simulated for different operating points giving an indication of the power generation capacity. Performance can be estimated at an early stage and machine design for prototypes determined. The simulations the effects of the blade contour on entry and exit at the free surface can be resolved well if the flow is modelled with another degree of freedom. Here, contours of pressure and torque would be able to identify the downstream regions of the blades in which modifications of blade geometry could improve flow conditions. This approach implies redefining the 3D model domain with the related mesh refinement.

In view of the above mentioned topics, it can be expected that a 3 dimensional model will be able to address these issues. This model will have a much larger domain and will require large computational capacity. The 3D model will also be able to implement curved blades or blade profiles. In general, the results of the 3D model can provide more accurate results. Further, the factors influencing the losses can be modelled thus providing for a complete model of the machine in the channel.

Chapter 6
Application of 3D CFD modelling

The modelling of the machine was done in two steps. Since flow conditions within a channel are complex in 3-D, the initial step consisted of setting up a 2D model for the wheel with straight blades as described in Chapter 5. Here the processes involved in the filling of the chambers could be explained. In a further step, the 3D wheel was modelled in a channel to enable the implementation of asymmetrical elements in the machine geometry (e.g. curved blades). In consideration of the free water surface and machine geometry with all features, the whole machine was modelled within the channel. Based on the results of the 2D model, the machine model was built in 3 dimensions, enabling mass flow in the transverse direction (across the width of the flume). This addressed issues raised in the limitations of the 2D model with regard to high discharges and the filling and emptying of the chambers of the rotor. It also enabled simulations with asymmetric modifications in blade geometry, thereby enabling an improvement of the blade geometry without regard to symmetry. For the 3D model, the domain was extended to include the narrowing inlet vent upstream and the outlet downstream of the rotor based on the CAD data from the experimental flume at the University of Darmstadt. The dimensions extracted from the CAD plan to model the computational domain are shown in the Figure 6-1. The profile of the flume bed includes the bottom step on the inlet side (inlet duct to channel bed) and the step below the rotor. In the initial domain, a half-model is mirrored at a plane in mid channel to reduce computational time.

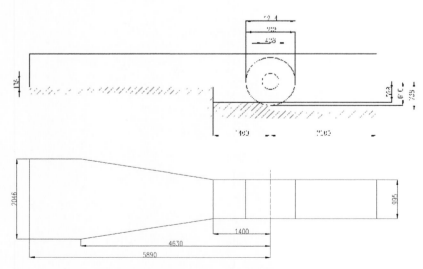

Figure 6-1: Dimensions of the flume (in millimetres)

6.1 3D Setup: Preprocessing

6.1.1 Domain

The complete machine domain consisted of two parts: the region of the wheel with blades and the region of the surrounding flume. The blade geometry was constructed and meshed separately to enable interchangeability of the wheel blades such that the entire domain need not be re-meshed for changes in blade geometry. The geometry of the rotor with 12 blades was constructed using data provided by the experimental set up. The following sequence (Figure 6-3) shows the two views of the construction and meshing of the central part of the CAD model. The region representing the central section of the flume (channel) around the rotor is defined by a set of points, lines and surfaces before being meshed. Figure 6-2 (a) shows the flume outline and in Figure 6-2(b), the volume to be meshed is shown in blue. The gap between the rotor and flume wall is not modelled for the initial run.

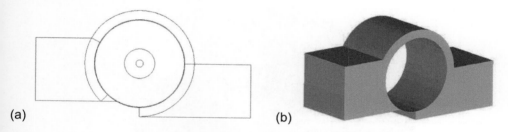

(a)   (b)

Figure 6-2: Central part of the domain

6.1.2 Mesh

The machine hub and blades were defined by a set of points and lines which describe hub diameter, blade diameter and blade thickness as well as the breadth of the machine (Figure 6-3 (a)). Using these features, surfaces were created. Subsequently the fluid region was defined and the mesh was generated. The outer limits of the mesh around the rotor were set by the cylindrical surface surrounding the blades and its sides (Figure 6-3 (b). The mesh of the rotor with straight blades in the computational domain was replaced by the mesh with modified blades for full 3D simulations (Figure 6-3 (c)). The asymmetrical blades show that a 3D model is required to fully represent the machine.

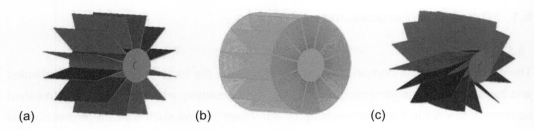

(a)   (b)   (c)

Figure 6-3: Blade construction and mesh

The computational mesh was divided into different parts to enable local refinement. Figure 6-4 shows the meshed wheel within the computational domain. As mentioned above, for the initial 3D run of the rotor with straight blades, the mid-plane of the flume was defined by a symmetry boundary condition.

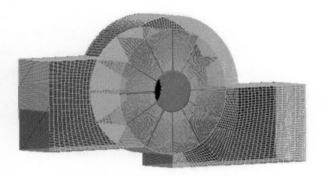

Figure 6-4: Mesh with straight blades

In a further step, the above mesh was extended to include the tapered inlet duct, thus enabling an increase in flow velocity leading up to the rotor. The outlet was also extended to include the downstream portion of the flume. The extension of the outlet channel can enable a better representation of flow conditions as found in the flume. The complete model with inlet duct, channel, rotor and extended outlet is shown in Figure 6-5 with the plane at mid channel representing a symmetry boundary condition.

Figure 6-5: Model with extended domain

In a further development of the model, the wall gap at the left wall was included. The model with the symmetry plane can gave an estimation of run time requirements. As the symmetry plane in mid-channel was not appropriate for the rotor with asymmetric blades, the full model included the entire width of the flume with wall gaps (Figure 6-6) at a later stage. This resulted in an increase in mesh size and computational time.

Figure 6-6: Full domain of the model

6.1.3 General settings

General settings were identical to those of the 2D models consisting of the pressure based VOF solver, explicit model with open channel flow, different cell zones for stationary domain and rotating mesh. Boundary conditions for the model were also identical to those of the 2D runs using a pressure inlet and pressure outlet as prescribed by the VOF model.

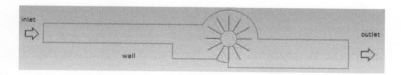

Figure 6-7: Boundary conditions

In the 3D case all boundaries are represented by surfaces (inlet, outlet, blades, walls). The geometry of the runner with straight blades enables the definition of symmetry (mirror) along the middle of the channel. The initial model was in reality a half-model with a symmetry plane representing the other half, thereby reducing the size of the model and computational time of the simulations. The domain extends to 4.6 m upstream and 3 m downstream of the shaft of the rotor. The model consists of 540000 elements.

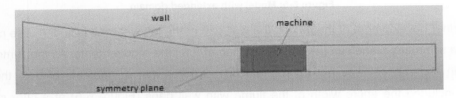

Figure 6-8: Top view of computational domain

In preparation for the runs with asymmetric blade geometry, the domain was extended in width to include the right channel wall along with wall gaps between wheel and channel walls (Figure 6-9) to enable a true representation of the machine in the experimental flume. The shaft of the rotor was extended up to the walls of the channel.

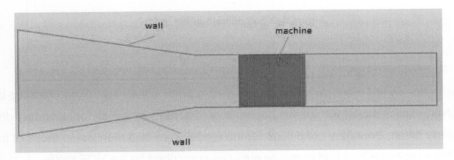

Figure 6-9: Full model

6.1.4 3D Cases

Four cases were set up with the numerical model. For the initial runs, Case 1 from the 2D set up with low discharge and low rpm were chosen (5.2.1 Case 1). The domain consisted of the upstream and downstream regions as defined in the mesh of the 2D models. Modifications in the geometry and mesh were made in subsequent runs when an increase in efficiency was seen.

In Case 2 the modifications in the geometry were made to adapt to the experiment under real conditions. The effects of these changes on the flow are visualised in the analysis.
For Case 2 a higher discharge, rotational speed and head difference was chosen. Here the domain size was increased in three steps, leading to the development of the full model. The effects of the increase in domain size on the results were examined. For the validation, the blade geometry was modified for the simulations as done in the experiments. The new blade geometry was run with the full model.

Case 3 was set up with a different rotational speed and head difference. In a further run, modifications in the blade geometry were taken over from Case 2 under the new boundary conditions. Results were compared with those of straight blades and could be verified using experimental results.

Variations in channel geometry and their effects on the performance of the machine were examined in Case 4. Here too, visualisation of flow conditions were examined.

The cases modelled are shown in Table 6-1. As in the previous Chapter 5, water levels are measured from the axis of the hub.

Table 6-1: Case setup for 3D model simulations

	Discharge [l/s]	Upstream level h_{up} [m]	Downstream level h_{down} [m]	$\Delta h = h_{up} - h_{down}$ [m]	rpm [1/min]	
Case 1						
	60	0.010	-0.420	0.430	2.47	
	60	-0.002	-0.213	0.211	3.11	
Case 2						
initial domain	180	0.203	-0.416	0.619	8.78	
symmetry plane	180	0.203	-0.416	0.619	8.78	
full model	180	0.203	-0.416	0.619	8.78	
full model, 10° blades	180	0.203	-0.416	0.619	8.78	
full model, 20° blades	180	0.203	-0.416	0.619	8.78	

	Discharge [l/s]	Upstream level [m]	Downstream level [m]	Δ h= $h_{up} - h_{down}$ [m]	rpm [1/min]	
Case 3						
full model	180	0.303	-0.408	0.711	7.8	
full model, 20° blades	180	0.303	-0.408	0.711	7.8	
Case 4						
	90	0.203	-0.416	0.619	8.78	

Modifications:

- shortening domain
- triangular obstructions
- variation in blade width
- bed slope
- wall gaps
- downstream diffusor

6.2 3D Analysis

6.2.1 Case 1

Two runs were set up in Case 1 (discharge 60 l/s) with a head differences of $\Delta h = 0.211$ m
and $\Delta h = 0.430$ m with rotational speeds of 3.11 rpm and 2.47 rpm respectively. The model
for the first run consisted of half the width of a rotor with straight blades. Symmetry
conditions were placed on the walls of the channel without accounting for wall gap losses in
the flow similar to the boundary conditions imposed on the 2D model. A case set-up with
low discharge (60 l/s) and head difference (0.211 m) was chosen.

The torque generated by the 2D model is shown in the following Figure 6-10. As described
in the 2D simulations, the repetition of the series pf peaks and troughs are caused by water
entering and exiting the chambers of the rotor. These peaks are further influenced by wave
build-up on the upstream and downstream side.

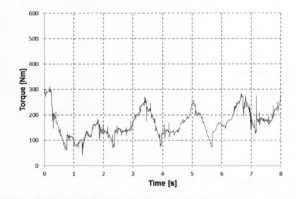

Figure 6-10: Torque generated by 2D simulation (Case 1, $\Delta h = 0.211$ m)

These peaks and troughs were also observed in the torque generated by the 3D model with
the straight blades and can be seen in Figure 6-11. The curve for the 3D run is smoother
than the 2D run indicating a better filling and emptying of the chambers.

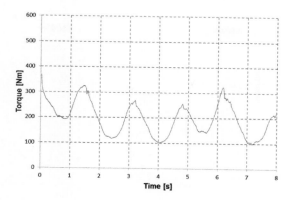

Figure 6-11: Torque generated by 3D simulation (Δh = 0.211 m)

For the 2D case with a larger head difference, the torque curve with strong fluctuation is shown in Figure 6-12. As described in the 2D case the variations move around a mean value.

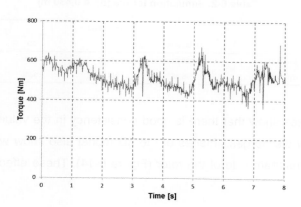

Figure 6-12: Torque generated by 2D simulation (Case 1, Δh = 0.430 m)

As seen in the previous 3D case, the curve for torque in the simulation with a larger head difference also shows the same characteristic. Figure 6-13 shows the curve for the 3D simulation with a head difference of 0.43 m.

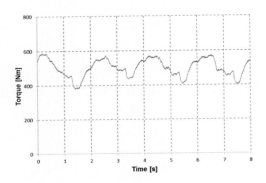

Figure 6-13: Torque generated by 3D simulation (Δh = 0.430 m)

Table 6-2 shows a comparison of the mean values for torque calculated from the above runs in 2D and 3D respectively. Here too the differences in the values for torque are nominal.

Table 6-2: Simulation torque (Δh = 0.430 m)

		Torque [Nm]	
	Measurement	2D	3D
Δh = 0.211 m	248.78	181.99	188.08
Δh = 0.430 m	539.45	497.55	502.15

The above two cases show that there is good consistency in the values obtained from the two models for low discharges. Images of the 3D model also show wave build up on the upstream and downstream side of the rotor (Figure 6-14). These effects were observed in the 2D simulations.

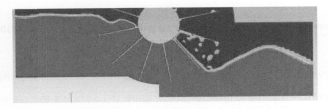

Figure 6-14: Wave on downstream side of the rotor

6.2.2 Case 2

To compensate for the effects of waves on the downstream side of the rotor for Case 2 (discharge 180 l/s, Δh = 0.619 m), the extent of the model was increased at the inlet and outlet to include the conditions of the experimental flume, maintaining the symmetry condition for the side walls. This change reduces wave build-up as shown in the contours representing the two phases water and air in Figure 6-15. The figure shows a wave entering the chamber as the blade is submerging. On the downstream side the tip of the blade of the chamber which has just emptied causes a wave on the water surface to move downstream

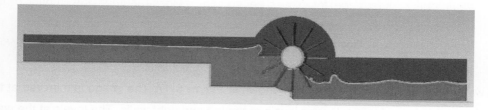

Figure 6-15: Extension of model

The processes involved in the filling and emptying of the chambers have been described in the 2D analysis (Chapter 5). On the basis of the 3D results, the conditions of flow at the wall gaps were analysed. The presence of the wall gap within the model should allow some of the water to escape downstream between the wheel and the walls. Further, the wall gaps can enable an easier emptying of the chambers.

Figure 6-16 shows the torque generated by the model without a wall gap. For the first three seconds the chambers on the downstream side do not empty and water is carried to the upstream side. Subsequently the amount of water carried upstream reduces, the torque increases and the values stabilise around the mean value of 310 Nm.

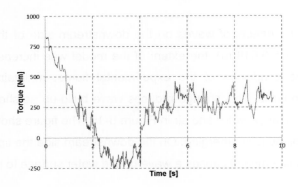

Figure 6-16: Torque, Case 2 without wall gap

The water level upstream of the rotor is shown in Figure 6-17. The image is mirrored at the mid-plane and shows identical water levels at the left and right walls. The levels at the walls are lower than at mid-channel as the mid-plane due to the symmetry condition imposed on the mid-plane.

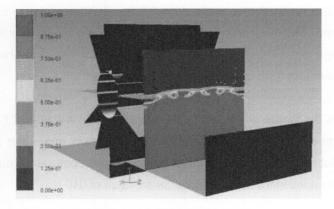

Figure 6-17: Water level before the rotor

Contours of velocity on the free surface are shown in Figure 6-18. The contours show an increase in velocity in flow direction in the inlet duct, reaching a maximum value at the narrowest point at the entrance to the channel. On the upstream side a high flow velocity at the free surface at the point of the blades entering the water. High flow velocities are seen with water exiting the chambers on the downstream side of the machine. This is also due to the vertical component of velocity of water flowing out of the upper chambers.

Figure 6-18: Surface velocity contours in m/s (discharge = 0.18 m³/s, Δh = 0.619 m)

High flow velocities are seen in the inlet duct. These velocities are maintained below the surface as the flow enters the channel from the inlet duct. When viewed from the symmetry plane of the model, low velocities are seen at the step (in blue) in the bed of the channel inlet indicating a zone of recirculation (Figure 6-19). High velocities are also observed at the entry of flow into the wheels chambers and at the gap between channel bed and blade tip. On the downstream side of the wheel high velocities are seen as the flow exits the chamber. The zone of high velocity in the flow at the blade tip near the step directly below the wheel is caused by the pressure difference between the closed chamber and the open chamber. Another zone of recirculation occurs at the step in the channel bed directly below the wheel (Figure 6-19).

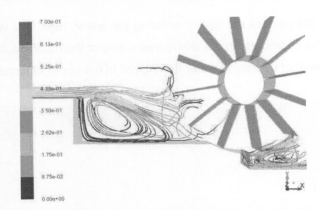

Figure 6-19: Velocity pathlines in the flow (discharge = 0.18 m³/s, Δh = 0.619 m)

The velocities within the zones of recirculation are very low compared to the surrounding regions, with the main flow passing over these regions.

6.2.3 Model with wall gap

In a further extension the wall gap was included in the model. This modification results in an increase of torque generated (Figure 6-20).

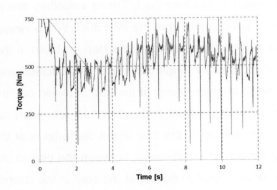

Figure 6-20: Torque, Case 2 with wall gap

The effects of the inclusion of the wall gap in the model are analysed below. On leaving the inlet duct and entering the channel, the flow has a region of low velocity in mid channel shown in Figure 6-21 (a). Through the narrowing of the inlet duct, the flow moves towards mid-channel (Figure 6-21 (b) thereby reducing velocities at the wall of the channel.

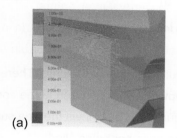   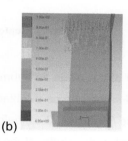

(a)   (b)

Figure 6-21: Velocity vectors at entry into channel

Figure 6-22 shows the water levels at the left wall (a) and on the mid plane (b) of the channel. Water is able to enter the chamber through the wall gap with the blade tip being submerged. In mid-channel the leading edge of the blade has not been submerged. While entering the free surface, the impact of the blade creates a wave, thereby displacing the water at the leading edge. This wave moves against the flow and hinders water from entering the chambers across the breadth of the blade (at the leading edge). At this moment, the water level in the wall gaps is higher than at the leading edge of the blade and water begins to enter the chamber at the wheel hub. Water which has entered the chamber through the gap has reached mid channel at the hub of the machine and begins to fill the chamber before the leading edge of the blade submerges. There is a drop in torque as the chamber is not filled and the tail water level is constant.

Within the wall gap at the hub of the machine, water flows to the downstream side accounting for hydraulic losses. Part of the flow passes over the shaft of the rotor (Figure 6-22 a)). On the downstream side the leading edge of the blade of the opening chamber is still below the downstream water level. Air entering the chamber through the wall gap enables the emptying. At mid channel the downstream water level has been reached in the chamber (Figure 6-22 b)).

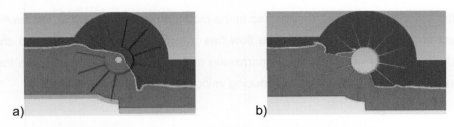

Figure 6-22: Contours of phases at a) left wall, b) symmetry plane

As a result of the blade tip submerging, a wave builds up in mid channel. There is a rise in water level in the gap as the blade forces water upwards. This is seen in the contour of the free surface (Figure 6-23 a)) before the wheel. Figure 6-23 b)) shows the free surface at the leading edge of the blade, at the gap and within the chamber being filled.

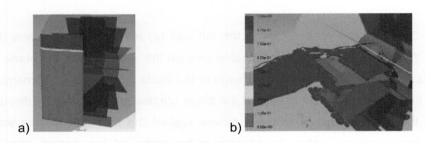

Figure 6-23: Free surface contours, a) before the blades, b) submerging blade

The formation of a wave hinders the flow from entering the chamber at the leading edge of the blade being submerged. Water entering the chamber through the wall gap begins to fill the chamber from the hub and flows towards the leading edge as the wheel rotates. Higher velocities are seen in the gap and in the region around the hub. Velocity contours in Figure 6-24 show a region of low speed at the blades leading edge caused by wave formation.

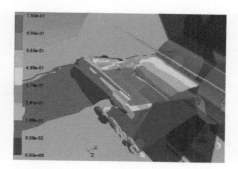

Figure 6-24: Velocity contours

As the chamber above the segment of the channel bed begins to close, velocities in the flow increase (Figure 6-25). For the chamber beginning to fill, velocities increase at the wave crest of the submerging blades leading edge.

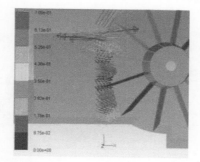

Figure 6-25: Velocity vectors before the rotor

The introduction of a wall gap between blades and channel walls show a change in the velocity distribution within the flow. At the symmetry plane in mid-channel (Figure 6-26) some recirculation occurs at the inlet step and at the submerging blade tips. Recirculation also occurs at the step below the wheel in mid-channel..

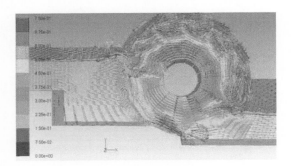

Figure 6-26: Velocity vectors, mid-channel (discharge = 0.18 m³/s, Δh = 0.619 m) with symmetry plane

The velocities at the channel wall show higher values than at mid channel indicating that flow conditions are more influenced by the wall gap than by the position of the blades with regard to the closing of the chambers (Figure 6-27). The zone at the step below the wheel in mid-channel also shows higher flow velocities.

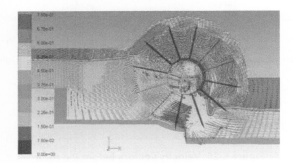

Figure 6-27: Flow velocities at left wall (discharge = 0.18 m³/s, Δh = 0.619 m) with symmetry plane

Path-lines of velocity in Figure 6-28 show a decrease in velocity as the flow enters the channel. On moving towards the wheel, velocities increase at the chamber being shut. On passing through the rotor higher velocities are caused by the emptying of the chambers. The step below the rotor causes recirculation within the flow with low flow velocities.

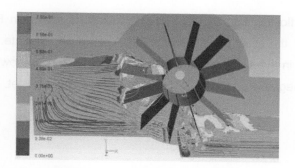

Figure 6-28: Velocity on the surface with path-lines in the flow

6.2.4 Full model with straight blades

Results of the simulation of the full model with straight blades show variations in the torque generated compared to 2D or 3D with a symmetry plane due to the effects of the blades entering the free surface. These are shown in Figure 6-29. The peaks and troughs in the course of the simulation are caused by blade impact, filling of the chambers and by downstream effects.

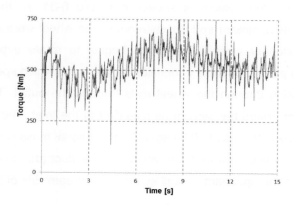

Figure 6-29: Torque - straight blades (Δh = 0.619 m)

The mass of water flowing into and out of the domain is shown in Figure 6-30. At the beginning of the run, the initial conditions for the water level dominate. As the simulations proceed, equilibrium in the system is attained. The positive series show water entering into the system and the negative series represent water exiting at the outlet.

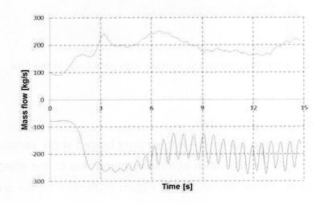

Figure 6-30: Mass flow rates

The mass balance for the system is shown in Figure 6-31. At the beginning of the simulation water levels upstream and downstream of the wheel were defined in the initial conditions for a stationary wheel. The wheel begins to rotate expelling water to the downstream side. As per the boundary condition for the outlet which specifies a given water level, the water passing through the wheel exits through the outlet. The upstream water level drops and water enters the system as specified at the inlet. The mean value for the mass balance in Figure 6-31 lies at 10.84 kg/s. This additional mass is distributed within the domain in the form of a rise in water level within the inlet duct (due to the narrowing of the flume) or as waves on the upstream as well as the downstream side of the rotor.

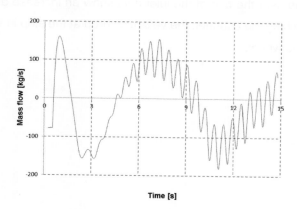

Figure 6-31: Mass balance (inflow + outflow)

The contour plot of velocity on the free surface is shown in Figure 6-32. The velocity contours show symmetry along the mid channel plane. As the inlet duct narrows, the surface velocity increases. A maximum on the upstream side is reached at the inlet of the channel leading to the wheel. High velocities at both walls of the channel and at the left and right sides of the blades can be seen in the contour plot. The downstream side shows highest velocities directly after and below the wheel. Figure 6-32 also shows that the symmetry boundary condition placed at mid-channel gives an accurate representation of the flow in the case of the machine with straight blades.

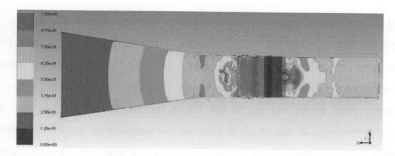

Figure 6-32: Velocity of the free surface (discharge = 0.18 m³/s, Δh = 0.619 m) full model

Velocities in mid flume near the bed of the inlet duct show an increase as the duct narrows (Figure 6-33). On reaching the channel, the velocities drop at the step in the bed, continuing to do so on nearing the wheel.

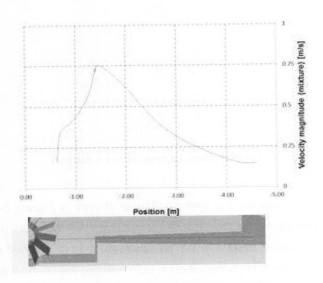

Figure 6-33: Velocity profile from the inlet to the blades

The flow conditions within the flume were analysed by plotting the velocity distribution in mid-channel and at a distance of 100 mm from the left wall.

Mid-channel

On entry into the channel, there is a reduction in the flow velocity at mid-channel (Figure 6-34). The regions coloured orange in the upper part of the figure represent velocities of the air phase. This region is separated from the liquid phase by the interface.

Velocity vectors show lower values in the liquid phase. Before the wheel, the velocity decreases due to the step in the channel bed. At the wheel, flow velocity decreases due to wave formation or the impact of the blades entering the channel. The change in velocity depends on the position of the blades with respect to time. Recirculation of flow occurs at the step in the channel.

High flow velocities are seen in the gap below the wheel. As in the 2D cases, the emptying of the chambers account for these velocities as the water flows out due to the pressure difference between the full chamber and the atmospheric pressure.

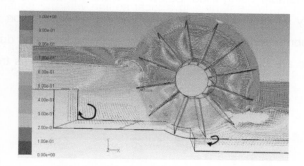

Figure 6-34: Velocity distribution at mid channel (discharge = 0.18 m³/s, Δh = 0.619 m) full model

Left wall

On nearing the hub, flow velocities in the channel increase in the region of the wall gap. Figure 6-35 shows the velocity distribution on a plane near (200 mm) the left wall. At this time step, no wave formation is seen as compared to the above Figure 6-34. Instead, the water surface is lower indicating that the water flows in the direction of the wall gap.

On the downstream side the chamber has begun to empty while the leading edge of the blade is still submerged in mid-channel (Figure 6-34).

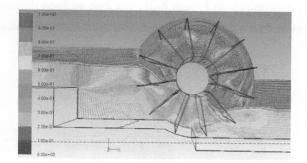

Figure 6-35: Velocity distribution at left wall (discharge = 0.18 m³/s, Δh = 0.619 m) full model

Recirculation at the upstream step and flow through the wall gap (right wall) are shown by the velocity path-lines in Figure 6-36, for the same time step as in the Figure 6-35previous figures. The figure shows path-lines of high velocity above the water surface caused by air escaping from the chamber being filled. Below the water surface the path-lines show high velocities upon entry into the channel, with considerable recirculation at the step in the channel bed. The submerged chambers also show some recirculation during the rotation of the wheel. Velocities are high for the fluid phase in the wall gaps as the flow passes the wheel.

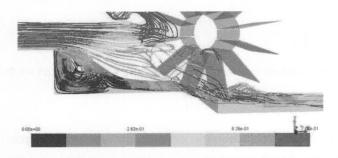

Figure 6-36: Velocity path-lines (discharge = 0.18 m³/s, Δh = 0.619 m) full model

The influence of the wall gaps on the flow upstream of the wheel show that velocities increase towards the sides of the channel. This indicates that the water level upstream of the rotor should be lower at the walls than in mid channel. This difference in water levels is in the range of a few millimetres. In Figure 6-37, the free surface directly before the wheel shows a drop at the walls of the channel indicating the formation of a wave in mid channel.

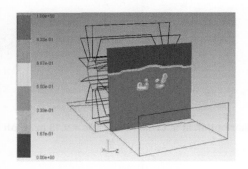

Figure 6-37: Water level before wheel (discharge = 0.18 m³/s, Δh = 0.619 m, full model)

As the wheel geometry is symmetrical along the shaft, the distribution of pressure is too, with regions of high pressure on the corners of the blades (Figure 6-38).

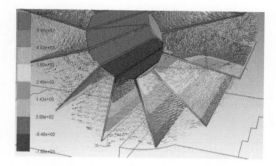

Figure 6-38: Pressure contours (blades), velocity vectors

6.2.5 Full model with blade slope

In the case of straight blades, the entry of the leading edge along the whole breadth of the wheel caused wave formation on the upstream side, resulting in a loss of performance. To counter this effect, the blades were angled at 20° to the axis to enable better conditions on entry of the blades into the free surface. This alteration enables the blade to enter the free surface at one point and allows the entire leading edge to be submerged gradually in the course of rotation of the wheel. Figure 6-39 shows the rotor with angled blades and with initial water levels. Additionally, the angled blades together with the curved segment of the channel bed enable the emptying of two chambers simultaneously.

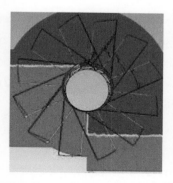

Figure 6-39: Rotor with angled blades

The following figures show the torque and mass balance from simulations with the rotor with 20° blades of the simulation with the full model.

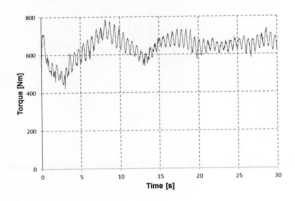

Figure 6-40: Torque (Δh = 0.619 m)

As the resulting torque on the shaft fluctuates with the filling and emptying of the chambers, the mass flowing into the domain also varies. This occurs at the outlet, with mass leaving the domain. Mass balance is thereby maintained with the difference between inflow and outflow fluctuating around zero.

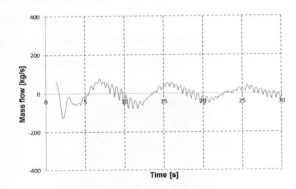

Figure 6-41: Mass balance

The change in blade angle and the subsequent entry point of the blades in the free surface at the left wall of the channel influence the flow upstream of the wheel creating higher velocities along the left wall (Figure 6-42). In consequence, there is a reduction in flow velocity in the right half of the channel before the wheel. On the downstream side, higher velocities are also seen on the free surface at the left wall as the chambers on this side open first. The 20° blade angle involves two chambers simultaneously in the filling and emptying process as the wheel rotates.

Figure 6-42: Velocity of the free surface (discharge = 0.18 m³/s, Δh = 0.619 m, 20° blade)

Velocity vectors in three sections (left wall, mid-plane, right wall) of the channel are shown in Figure 6-43. At this time step, two chambers are partially open at the downstream step below the wheel. In the vector plot 200 mm from the left wall, (Figure 6-43, a)), there is some recirculation at the upstream step. Flow speeds below the surface are high at the wheel. After passing through the wheel, flow speeds increase as Chamber A (and the previous chamber) empty. Chamber B is partially open and allows water to pass through the wheel from the upstream to the downstream side. This results in high flow velocities at the step below the wheel which hinder recirculation. High flow velocities are seen at the blade tip which has opened chamber B to the downstream side of the channel. At mid-channel, this outflow however does not prevent recirculation at the step in the bed (Figure 6-43, b)). Flow velocities are lower at this plane. Figure 6-43, c) shows velocity vectors in a section 200 mm from the right wall. High velocities are found at the rotor in this region as the flow changes direction and moves towards the left wall in passing through the partially open chamber B. The wall gap allows regions of high velocities to form.

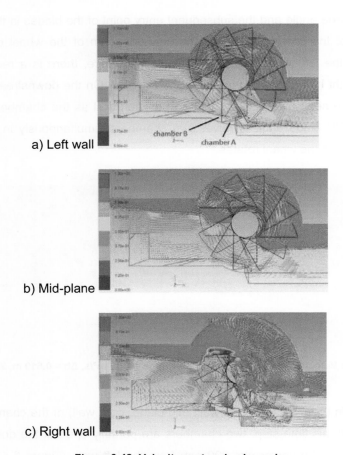

a) Left wall

b) Mid-plane

c) Right wall

Figure 6-43: Velocity vectors in channel

Figure 6-44 shows a partially submerged blade in the free surface upstream. As the blade moves downwards, the chamber begins to fill with water entering from the wall gap at the left wall and moving to the right. As seen in the simulation of the model with straight blades with a wall gap at the sides, higher flow velocities at the hub within the chamber influence the filling of the chambers.

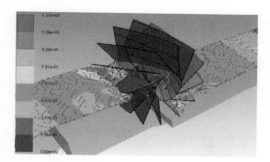

Figure 6-44: Blade entry into free surface upstream

These higher velocities are also observed below the free surface at the submerged tip (Figure 6-45) with the highest velocities in the region of the (left) wall. The influence of the asymmetric blades is also seen at the inlet above the step at the inlet.

Figure 6-45: Velocity contours and vectors

As the chamber fills, the flow forces the wheel to rotate creating regions of high pressure on the blades. As opposed to the straight blades where the pressure profile is symmetrical, the angled blades show maximum values at the leading tips on the left hand side of the wheel (Figure 6-46). This leads to an unsymmetrical load on the shaft of the machine.

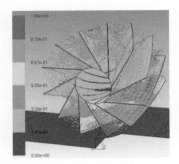

Figure 6-46: Blades, pressure distribution, velocity vectors in flow

Velocity path-lines show high values within the chamber being filled. When viewed from the right wall (Figure 6-47) high flow velocities can be seen within this chamber which is open to the upstream and downstream side of the channel. This increase in flow velocity before the chamber is caused by the shutting of the chamber. Within the chamber, velocities are additionally influenced by the step in the channel bed. As a result, recirculation occurs locally.

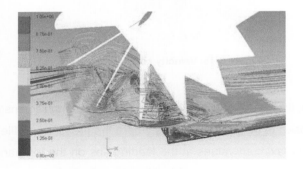

Figure 6-47: Velocity path-lines, right wall

As the chamber is shut by the segment at the left wall, the flow can only move downstream by passing through the wall gap and the gap between the blades and the channel bed. Velocity path-lines in Figure 6-48 show low values at the blade tip thereby creating a high pressure on the blade. Flow velocities within the chambers opening out onto the downstream side reach a maximum at the step in the channel bed. There is an increase in flow speed within a partially opened chamber.

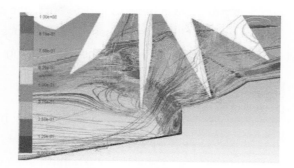

Figure 6-48: Velocity path-lines, left wall

On the downstream side, flow conditions at the step in the channel bed and below the rotor are determined by the opening of the chamber. In Figure 6-49 velocity is high at the left wall. The path-lines show that this increase also occurs below the free surface.

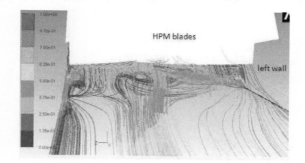

Figure 6-49: Velocity vectors downstream, upstream view

Through the asymmetry of the blades of the wheel, the main path of the flow is diverted to the left wall of the channel on the downstream side. Low velocities are seen at the right wall of the channel.

6.2.6 Case 3

As in the previous case, the effect of changes in the blade geometry was examined for a discharge of 180 l/s, a larger head difference ($\Delta h = 0.711$ m) and a lower rotational speed. Figure 6-50 shows the torque generated for the run with straight blades.

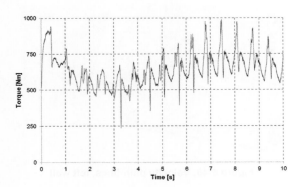

Figure 6-50: Case 3, straight blades

The second run was set up with blades set at an angle of 20° to the axis. The following Figure 6-51 show the values for torque for the simulation with a head difference Δh = 0.711 m with angled blades. The results show an improvement in the performance in the simulation.

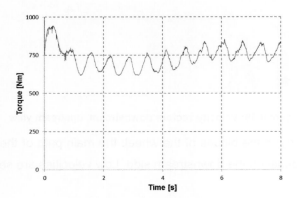

Figure 6-51: Torque for 20° blades, (Δh = 0.711 m)

The detailed analysis for the simulation time t = 2 s to t = 4 s show a fluctuation in torque from 650 Nm to 700 Nm as will be discussed in 6.3 Comparison of results.

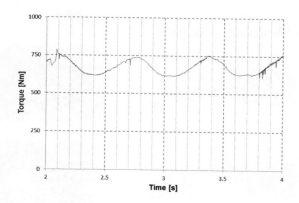

Figure 6-52: Torque on rotor (2 to 4 s)

The following sequence shows the velocities at the free surface of the flow in the flume (left column) and the free surface in the middle of the flume (Figure 6-53). At the time t = 2.00 seconds the leading blade tip is submerged and water is entering the next chamber shown by contours of high velocity (in red, at the left wall). Due to the angled blades, water is being expelled from the chamber by the sinking blade thereby showing high velocities near the right wall gap. As water is also being forced against the flow direction, a region with low flow velocity develops near the right wall upstream (coloured blue). The chamber below the leading blade tip is filled up to mid-channel (image, right column) while on the downstream side the chamber at the free surface is emptying.

As the leading edge of the blade submerges, velocities on the upstream side are high due to simultaneous inflow and outflow (t= 2.10 s to 2.40 s). During this time, the chamber downstream is still filled. While the chambers on the upstream side begin to fill from the leading blade tip at the left wall, the emptying of the chambers downstream begin at the right wall. Water flows out of the chamber under the influence of gravity, leading to a drop in velocities at the free surface. At t = 2.40 s there is a drop in torque as the flow strikes the hub and changes direction in the upstream chamber. The downstream chamber is partially filled, with high outflow near the left wall. As the upstream chamber fills whilst the downstream chamber empties, torque increases (t = 2.50 s).

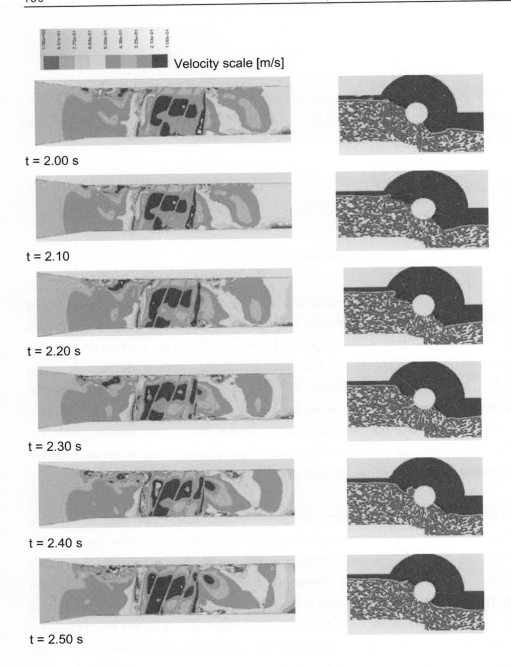

Velocity scale [m/s]

t = 2.00 s

t = 2.10

t = 2.20 s

t = 2.30 s

t = 2.40 s

t = 2.50 s

Figure 6-53: Free surface velocities, air-water phases (t = 2s to2.5 s)

The velocities at the left and right walls upstream are caused by the interaction between the wheel blades, the V-shaped inlet duct and the step in the channel bed at the end of the duct. The following sequence (Figure 6-54) shows a region of low velocity spreading upstream at the right wall, initiated by the closing of the chambers against the free surface with water being expelled, subsequently causing a wave to propagate upstream (t = 2.60). (Flow velocity at the left wall increases as water enters the chambers and the chambers begin to fill.) A peak in the torque at t = 2.80 is caused by the filled upstream chamber and the low water level on the downstream side. As the next downstream chamber does not empty at t = 2.90 and t = 3.00, torque decreases.

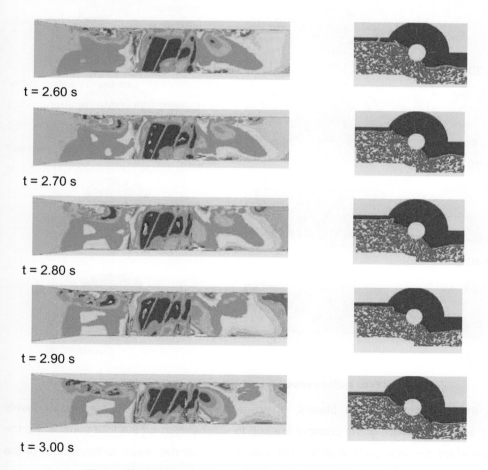

t = 2.60 s

t = 2.70 s

t = 2.80 s

t = 2.90 s

t = 3.00 s

Figure 6-54: Free surface velocities, air-water phases (t = 2.6s to 3.0s)

A further cycle for the increase in torque in Figure 6-55 (t = 3.1s to 3.5s) show the effects of the filling of the chambers upstream. On the downstream side, flow velocities increase with the emptying of the chambers and a corresponding drop in water levels at the blade.

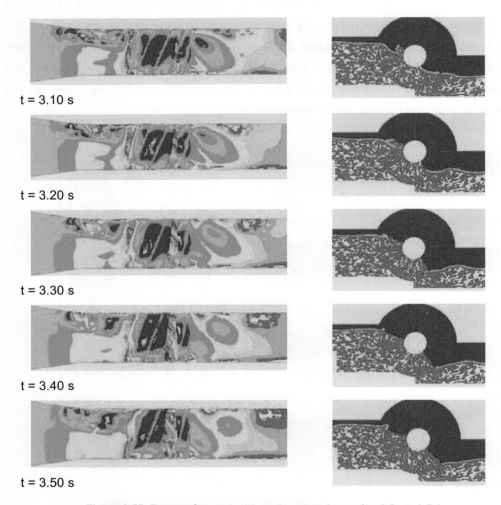

t = 3.10 s

t = 3.20 s

t = 3.30 s

t = 3.40 s

t = 3.50 s

Figure 6-55: Free surface velocities, air-water phases (t = 3.5s to 3.5s)

Due to the asymmetry of the blades, upstream effects include different water levels (waves), across the width of the channel leading to recirculation on the free surface. As the blade tips enter the free surface at the left wall of the channel, water is forced out of the chambers at the right wall. Additionally, the effect of the step in the channel bed influences recirculation. Vortices form in the upstream channel and cause turbulence on the free surface as well as within the flow. The velocities and water levels which lead to a build-up in torque are shown in Figure 6-56 from t = 3.60 s to 3.40 s.

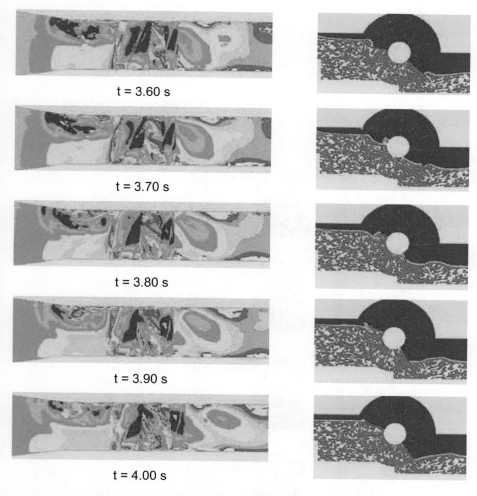

t = 3.60 s

t = 3.70 s

t = 3.80 s

t = 3.90 s

t = 4.00 s

Figure 6-56: Free surface velocities, air-water phases (t = 3.6s to 4.0s)

Similarly, the following Figure 6-57 shows the pressure distribution at the rotor together with the air-water phases and the velocity contours on the free surface. As such, the velocity distribution on the free surface does not play a significant role in torque generation but the effects of recirculation dictated by the blade geometry contribute to turbulence effects in the horizontal plane before the wheel.

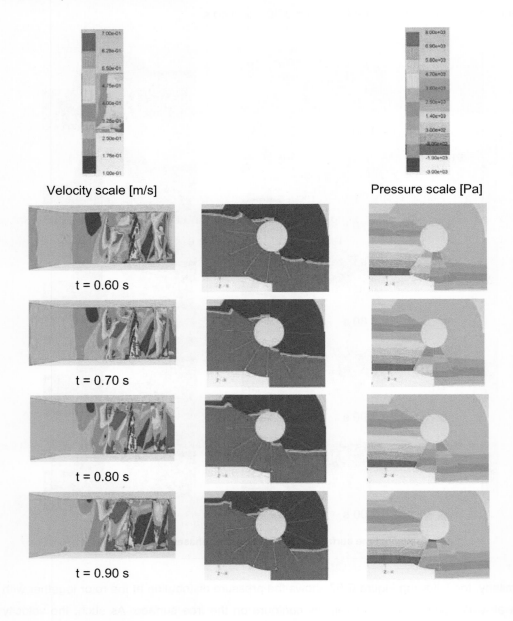

Velocity scale [m/s]   Pressure scale [Pa]

t = 0.60 s

t = 0.70 s

t = 0.80 s

t = 0.90 s

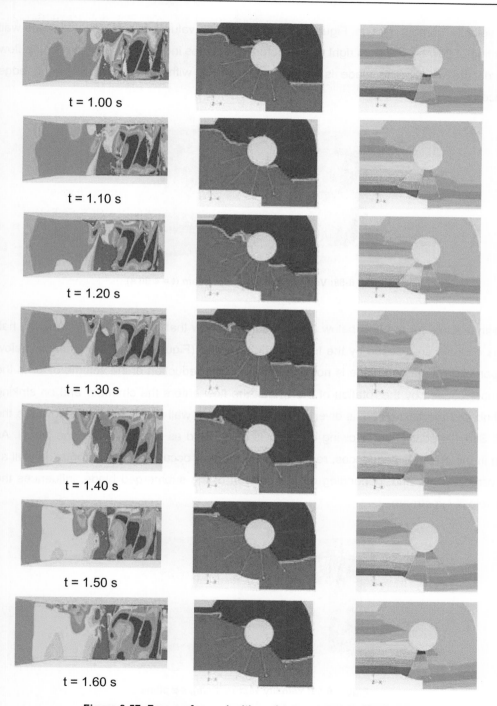

t = 1.00 s

t = 1.10 s

t = 1.20 s

t = 1.30 s

t = 1.40 s

t = 1.50 s

t = 1.60 s

Figure 6-57: Free surface velocities, phases, pressure distribution

The velocity contour plot in Figure 6-58 shows high values at the left and right wall (direction of flow here is from right to left) before the blades indicating a deviation in the flow towards the wheel. The blade is partially submerged, with a part of the leading edge exposed.

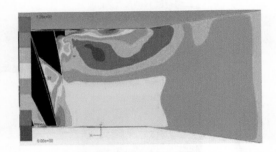

Figure 6-58: Velocity contours, upstream (t = 4.00 s)

Velocity vectors for the flow below the free surface show that high velocities in the left half of the channel are caused by the filling of the chamber (Figure 6-59). The chamber below the partially submerged blade is not filled. Through the reduction of the volume of air in the chamber caused by the rotation of the wheel, the flow enters the chamber and on striking the hub and the next blade, is diverted towards the right wall and is forced out towards the inlet. Simultaneously the following chamber is being filled as the flow drives the wheel. As seen in the previous sequences, recirculation in the flow occurs in the horizontal as well as the vertical plane and, depending on the position of the submerged blade, influences the flow velocity upstream.

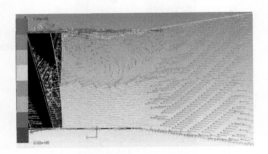

Figure 6-59: Velocity vectors [m/s], x-z plane

When viewed in the vertical (x-y) plane, the velocity vectors show that recirculation occurs at the step in the channel bed (Figure 6-60). The wave created by the preceding trailing blade tip moves upstream and is in the inlet duct. At the step in the channel on the inlet side, the combination of higher velocities in the flow entering the channel from the converging inlet duct an water being forced out of the chambers against the flow direction lead to recirculation at the step. Velocities directly before the blades show high values near the surface as the flow enters the chambers.

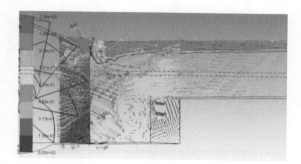

Figure 6-60: Velocity vectors with surface contour, x-y plane

The water level directly upstream of the blades is shown as contours of phases in Figure 6-61. The asymmetries of the blades cause sloshing within the channel directly before the wheel, resulting in vortices.

Figure 6-61: Water level, upstream

In Figure 6-62 a), path-lines of velocity show that as the flow enters the channel from the inlet duct, there is a strong zone of recirculation at the step on the inlet side of the wheel. Figure 6-62 b) shows that the recirculation has higher velocities at the right wall. The recirculation moves towards the left wall of at the step and thereby influencing the velocities within the flume.

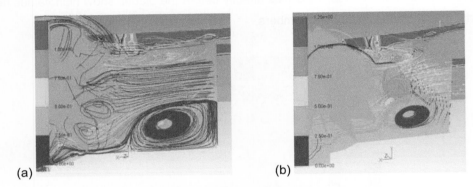

(a)   (b)

Figure 6-62: Recirculation at step on the upstream side

The zones of recirculation vary in size depending on the position of the blades and are influenced by the vortices built up at the surface. These vortices are drawn down by the submerging blades.

Forces acting on the rotor

The wheel was made up of two zones. One zone was a single blade (wall_blade_one), with the rest of the blades being contained in the other (wall_blades). Monitors of pressure and viscous forces action on the wheel are shown in Table 6-3. The table shows the forces at a single time-step. The vectors of force along the three axes as well as the resultant forces show that the viscous part of the forces is very small compared to the pressure forces.

Table 6-3: Pressure and viscous forces

Forces	Pressure	Viscous	Total
Direction Vector (1 0 0)			
Zone			
wall_blade_one	256.329	0.051	256.380
wall_blades	3294.288	0.913	3295.201
Net	3550.617	0.963	3551.580
Direction Vector (0 1 0)			
Zone			
wall_blade_one	704.038	0.461	704.499
wall_blades	-1253.690	-0.834	-1254.525
Net	-549.652	-0.374	-550.025
Direction Vector (0 0 1)			
Zone			
wall_blade_one	-273.471	1.219	-272.252
wall_blades	974.171	-0.102	974.069
Net	700.700	1.117	701.817
Direction Vector (1 1 1)			
Zone			
wall_blade_one	396.580	0.999	397.579
wall_blades	1740.578	-0.013	1740.564
Net	2137.157	0.986	2138.143

Forces acting on one blade of the rotor for one complete passage through the channel are shown in Figure 6-63. Two peaks are seen in the total force acting on one blade of the rotor. The first peak occurs when the blade tip enters the free surface, at the moment when the blade generates a negative torque. The second peak occurs when the blade closes its preceding chamber at the bottom of the channel.

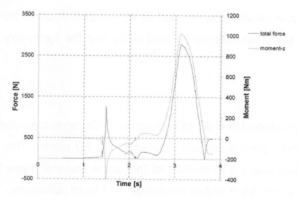

Figure 6-63: Total force acting on one blade

The forces acting on the blades will have to be considered during the construction as the strength of materials used will determine the endurance of the installation.

6.3 Comparison of results

Case 1

For the Case 1 the results of the 2D and 3D simulations showed similar efficiencies with η being the ratio of measured/simulated power at the shaft divided by the theoretically power available. The simulations showed an under-prediction with values lying below those of the measurements (Figure 6-64). In the case of the higher head difference, the simulation results are closer to the experiments.

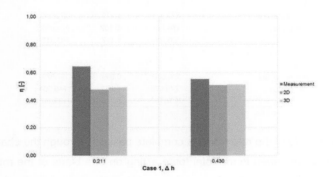

Figure 6-64: Comparison of efficiencies in Case 1

The lower values in the simulations are due to the simplification of the model through the omission of the wall gaps between wheel and walls and the symmetry boundary condition at mid channel.

Case 2

The 3D model was developed in a set of simulations with increasing domain size, beginning with the central channel portion as was defined in the 2D cases. The addition of features like wall gap, inlet duct and the extension of the outlet channel show that the simulations can be set up successfully for a higher discharge and head difference. The initial simplified model without any features as in Case 1 showed an underestimation in the performance (Figure 6-65, left). The addition of the wall gaps showed an improvement in the efficiency with results higher than those calculated from the experiment. The wall gaps allow for transverse flow during the filling of the chambers.

The results of the full model showed a further rise in the performance. With the inlet and outlet at a distance from the shaft of the wheel, friction losses in the flume could account for the lower experimental values.

Alteration in the blade slope angle showed a further increased in efficiency of the model. With the 20° blade angle, the model improved accordingly in comparison to the 10° blade angle measurements (Figure 6-65).

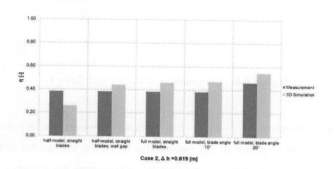

Figure 6-65: Addition of features in the simulation models

The simulation models do not consider mechanical losses which occur during the operation in the measurements. In reality these losses are accounted for e.g. by the bearings and braking system.

Case 3

As in Case 2, a mean discharge of 0.18 m³/s was chosen for Case 3 the head difference being 0.711 m. Identical boundary conditions imposed on the modified blade geometry models show an increase in torque generated on the model for Case 2 (Δh = 0.619 m) and Case 3 (Δh = 0.711 m). Figure 6-66 shows a comparison of the efficiencies for straight and angled blades in Case 2 and Case 3. In both the cases, measurements in the flume with angled blades were made with slightly higher head differences.

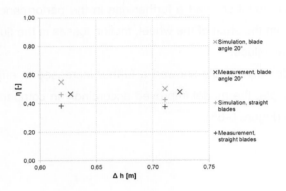

Figure 6-66: Comparison of Case 2 and Case 3

In Table 6-4 the difference in values for torque between experimental results and simulations is expressed in terms of fall head. The negative values for fall head indicate a decrease in water level and vice versa. These effects can be caused by wave effects through the narrowing of the inlet duct and the wave build up on the upstream and downstream side. The boundary conditions of water levels for the inlet and the outlet of the model were based on mean values measured during the experiments. On the inlet side, the values were measured in a part of the channel not included in the model domain. As a result, the effect of the narrowing of the inlet duct on the water level could lead to the increase in torque.

Table 6-4: Difference in torque in terms of fall head

$\Delta h = h_{up} - h_{down}$ [m]	Torque ($M_{expt} - M_{sim}$) [Nm]	Power (M_{sim}) [W]	Δh (from M_{sim}) [m]
0.211	62.59	20.38	0.034
0.430	37.29	9.65	0.017
0.619	148.73	136.69	0.077
0.619	-65.55	-60.24	-0.034
0.619	-95.30	-87.59	-0.049
0.619	-197.26	-181.29	-0.102
0.711	-191.02	-155.98	-0.090
0.711	-75.34	-61.52	-0.035

The differences in water levels lie in the range of a few centimetres. Considering the hydraulic and mechanical losses in the experiment which are not represented in the model, the results obtained are sufficiently accurate to predict the behaviour of the machine quantitatively.

6.4 Case 4

The previous simulations show that prediction of efficiency is close to those of the measurements. In the following case, variations in the channel geometry were modelled to examine the effects on the flow and on the performance. Case 4 was modelled for a discharge of 90 l/s and a head difference $\Delta h = 0.619$ m. The geometry of the model for Case 4 did not consider the long narrowing intake. The inlet was modelled above the step in the channel, upstream of the rotating wheel.

6.4.1 Influence of channel dimensions on machine performance

To reduce run time, the domain was reduced to that of Case 1 with the inlet located above the step on the upstream side of the wheel as in the experimental setup. Correspondingly, the outlet was placed on the channel bed downstream of the wheel. The channel-width, wheel-width, wall gaps and blade slope were maintained in the new domain. The results for the performance of the full model and the model with the shortened domain show a slight increase in values in Figure 6-67.

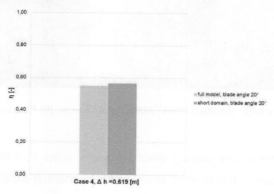

Figure 6-67: Short domain

As was seen in the Case 2, the filling of the chambers on the upstream side could be improved with the introduction of wall gaps, leading to a smoother curve for torque and a higher efficiency. Hence the wheel-width was reduced in comparison to the channel-width. In this case, small triangular obstructions were maintained on the sides of the wheel below the water surface. The alterations in the wheel dimensions were made based on the geometry of a downsized laboratory model at the University of Southampton having a breadth of 0.238 m. The simulation model was modified accordingly for a channel-width of 1 m and meshes were generated for wheel breadths of 0.8m and 0.5m. Wheel diameter, number of blades, blade thickness, and wall gaps at the struts were not modified. A plan view if the domain is shown in Figure 6-68 (a) with triangular obstructions (shaded in grey). The CAD model in Figure 6-68 (b) shows the inlet (green) and the step below it along with the triangular obstructions at the sides of the wheel.

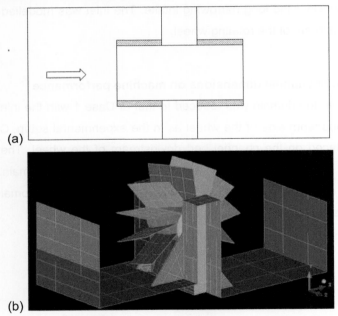

(a)

(b)

Figure 6-68: Modified wheel width (a) plan view, (b) CAD model

Results of the three runs with varying machine width show that the machine efficiency increases for a decrease in the ratio of wheel-width to channel-width. As described in the previous chapters, the recirculation of the flow before the wheel influences the performance of the machine. Reduction of the wheel breadth enables a better filling and emptying of the chambers. These modifications also show a change in flow conditions upstream of the wheel. The values for efficiency of the three runs indicate a peak performance for widths between 0.5 and 0.8 m (Figure 6-69).

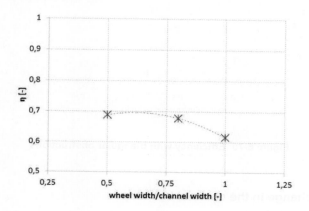

Figure 6-69: Efficiency with varying wheel width

A further reduction of wheel-width could show an improvement in the performance of the wheel. Simultaneously the narrower wheel would reduce absolute power generated. The predicted power generated by the model lies at around 350 watts.

6.4.2 Influence of triangular obstructions

The removal of the triangular obstructions at the sides of the wheel (upstream and downstream sides) shows a marginal change in the machine efficiency. Results of the simulations are shown in Figure 6-70.

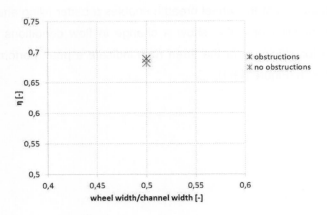

Figure 6-70: Efficiency with triangular obstructions

6.4.3 Effects of change in the channel bed slope

In the following figure, the region below the inlet shows a zone of recirculation in the channel. With the angled blade submerging, water is forced back towards the right wall and subsequently towards the inlet. Together with the recirculation below the inlet, the flow swirls towards the blades. Velocity path-lines in Figure 6-71 show this behaviour of the flow in the channel. The zone of recirculation at the step below the inlet has reduced. as a result of the angled blades a part of the flow has been diverted along the right wall.

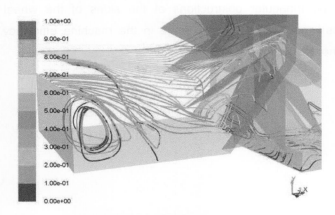

Figure 6-71: Velocity pathlines

To reduce the recirculation, the bed slopes within the channel were altered on the upstream and downstream sides of the rotor for a given discharge. Two different slopes were chosen for the upstream bed and three variations for the downstream side. Slopes chosen for the series are listed in the following Table 6-5.

Table 6-5: Channel bed slopes

Upstream slope	Downstream slope	Model
1:2	1:2	
1:2	1:5	
1:2	1:10	
1:3	1:2	
1:3	1:5	
1:3	1:10	

Figure 6-72 shows the efficiencies for the models with the chosen bed slope combinations. For a given upstream slope of 1:3 there is an increase in performance as the bed slope increases. An increase in the upstream bed slope show lower values for the corresponding changes in bed slope. The highest efficiency is obtained for an upstream bed slope of 1:3

and a downstream bed slope of 1:2. A further simulation with an upstream slope of 1:3 maintaining the original step below the wheel showed the least machine efficiency.

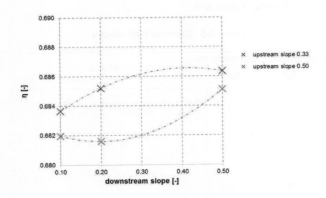

Figure 6-72: Bed slope variation

Simulations with further variations were performed on the model with the most efficient bed slopes 0.33 for the upstream slope and 0.5 for the downstream bed.

6.4.4 Effect of wall gaps

Using the model with the optimal bed slope, the effect of gap clearance between the wheel and the surrounding structure (channel walls and channel bed) were investigated. The resulting simulations showed high Courant numbers in the mesh at the wall gaps. These regions lie around the shaft of the wheel at the fluid-air interface. Due to high velocities in the air at the fluid-air interface of the VOF model, the cells at that region show high values for the Courant number during the emptying of the chambers. Local mesh refinement was carried out but the values for Courant number did not decrease sufficiently to complete the run. Figure 6-73 shows the Courant number in the wall gap with the free water surface. Here the chamber is still shut on the downstream side, forcing water through the narrow wall gap.

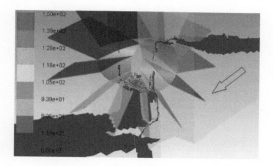

Figure 6-73: Courant number in wall gap

Hence the channel geometry and wheel dimensions were modified to eliminate the wall gaps completely. Two cases were modelled and run. In both cases, the elimination of the wall gaps showed an increase in performance. The change in overall efficiency is shown in Figure 6-74.

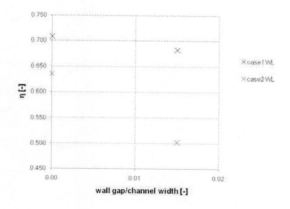

Figure 6-74: Variation in wall gap

Pressure distribution below the rotor

The effects of variation in the wall gap and downstream channel geometry on the pressure distribution below the rotor were monitored at three points at the hub and at the bed of the channel. Figure 6-75 shows the points 1 to 6 below the hub and on the channel bed.

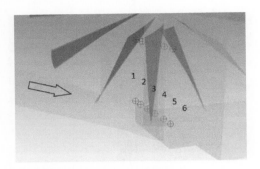

Figure 6-75: Pressure monitor points

Figure 6-76 and Figure 6-77 show the total pressure monitored at the points on the hub and at the channel bed below the wheel. The first figure shows the progression for total pressure calculated below the hub of the wheel. The pressure in mid-channel becomes negative as torque increases. At the peak in torque, the pressure below the hub shows a minimum.

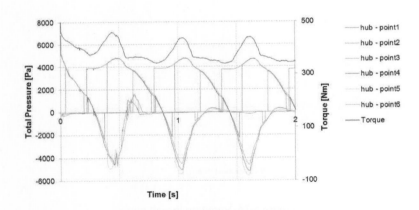

Figure 6-76: Pressure below wheel hub

In comparison to the pressure levels at the channel hub, the pressure monitors at the channel bed show positive values. A brief negative pressure is seen near the channel wall which is due to high velocities during the closure of the chamber.

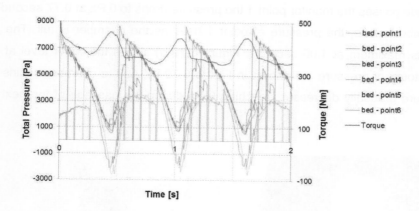

Figure 6-77: Pressure at channel bed

Figure 6-78 shows pressure distribution below the hub of the wheel at three points across the width of the channel with the values for the torque generated by the machine from 0.7 seconds to 1.4 seconds. At point 1 at the left wall of the channel the pressure does not drop below 0 Pa. At mid channel the pressure at point 3 below the hub varies between -3000 Pa and 5000 Pa. The values for generated torque reach a maximum when pressures below the wheel at mid-channel and at the right wall show minimum values (negative values).

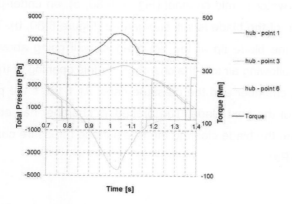

Figure 6-78: Pressure at wheel hub

As the blade passes the monitor point 1 the pressure drops to 0 Pa at 0.77 seconds. As the next chamber opens, the pressure at point 1 rises as the chamber is full. The pressure reaches its maximum at 1.06 seconds. Figure 6-79 a) shows the water level at this time step. Contours of pressure are shown in Figure 6-79 b). As the chamber begins to open, the pressure gradually decreases, reaching 0 Pa during the passage of the next blade at 1.35 s.

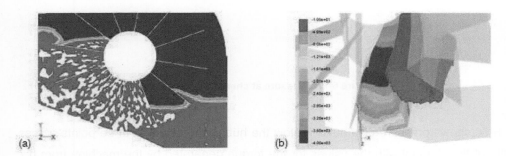

Figure 6-79: Point 1, 1.06 s, contours of a) density at mid channel, b) pressure

At point 3 in mid-channel the pressure drops at a constant rate as the chamber opens. The value reaches 0 Pa at 0.870 seconds. At 1.027 s the pressure below the hub at mid-channel reaches its minimum value. Although the contours of density show that the chamber still contains water at mid-channel (Figure 6-80, a), an under-pressure is created by as the leading edge of the blade is still submerged (Figure 6-80, b). The opening of the chamber begins with the blade tip at the left channel wall rising above the downstream water surface thereby allowing air into the chamber. As this opening of the chamber begins, water flows out of the chamber and the pressure rises. At 1.168 s the pressure reaches 0 Pa with the passage of the next blade. The preceding chamber has opened from the left wall to mid-channel with the blade clearing the free surface (pressure contours do not show pressures below -100 Pa).

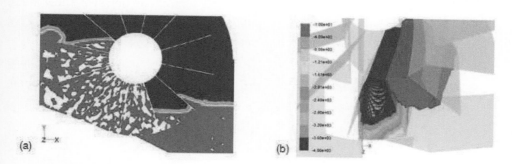

(a)

(b)

Figure 6-80: Point 3 1.027 s, contours of a) density at mid channel, b) pressure

At the right wall below the hub (point 6) the total pressure remains around 0 Pa, briefly dropping to - 793 Pa after the passage of the blade at 1.146 s. The sudden drop in pressure coincides with the passage of the blade as well as with the shutting of the chamber on the upstream side. The pressure continues to rise with the emptying of the chamber, reaching a positive value at 1.247 s.

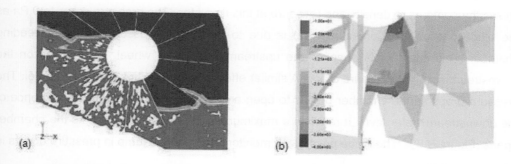

(a)

(b)

Figure 6-81: Point 6, 1.146 s contours of a) density at mid channel, b) pressure

The pressure distribution at the three points on the channel bed below the wheel is shown in Figure 6-82. Here the peak in torque corresponds with the drop in pressure along the channel bottom.

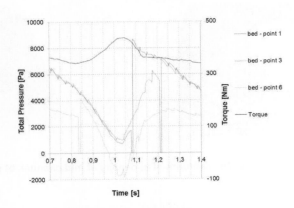

Figure 6-82: Pressure at channel bed

At point 1 near the left wall the pressure drops from 0.7 seconds as the chamber opens to the downstream side. At 1.028 seconds the pressure reaches its minimum value of 705 Pa and begins to rise due to the pressure of the downstream water column (level). Figure 6-80 shows the contours of density and pressure at this time step. The pressure drops to 0 Pa as the blade passes. The pressure jump is due to the higher pressure in the succeeding chamber. This chamber is open on the upstream side of the wheel. As it opens on the downstream side, the pressure drops. A similar effect is seen at point 3 in mid-channel. The pressure drops as the chamber begins to open and begins to rise through the influence of the downstream water level. It reaches a maximum at 1.1075 s and drops as the chamber opens (Figure 6-83). The blade passes the monitor point and the jump in pressure occurs in following channel.

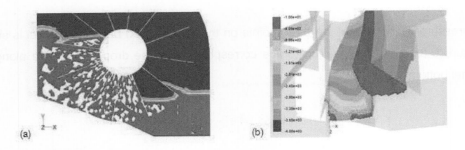

Figure 6-83: Point 3 1.076 s, contours of a) density at mid channel, b) pressure

At point 6 near the right wall of the channel pressure drops as the chamber empties. As the blade passes the monitored pressure drops to 0 Pa. The subsequent pressure jump is caused by the partially filled succeeding chamber and is not as high as at the other points monitored. At point 6 the total pressure is influenced by the downstream side of the rotor. The pressure drops below 0 Pa while the chamber is open on the downstream side. As the water level in the chamber rises, the pressure increases

6.4.5 Effect of downstream channel dimensions

In a further model the downstream side of the channel was modified by keeping the downstream channel-width equal to the width of the rotor (Figure 6-84).

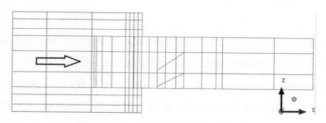

Figure 6-84: Downstream channel

Results of the simulation do not show a significant change in efficiency of the machine. In the case of the channel with an 8° diffusor on the downstream side of the rotor (Figure 6-85) the simulation shows a drop in the overall efficiency of the machine.

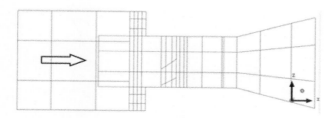

Figure 6-85: Diffusor

Results of the simulations for the model with modifications at the downstream of the channel are shown in Figure 6-86.

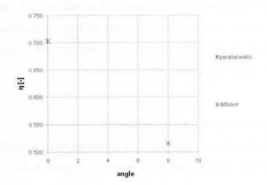

Figure 6-86: Efficiency, downstream modification

Contours of the phases (water levels) and velocity vectors of the simulation model with parallel channel walls downstream and a diffusor are shown in the following Figure 6-87 in the left and right column respectively. In Figure 6-87 a, the water level at on the downstream side are higher for the model with parallel walls than for the diffusor model Figure 6-87 b. The model with parallel walls shows higher velocities in the flow in the region towards the outlet. These velocities are visible in the vertical plane in mid-channel as well as in the horizontal plane near the channel bed Figure 6-87 c, e. The simulation of the model with the diffusor shows larger regions of low velocity in both planes Figure 6-87 d, e.

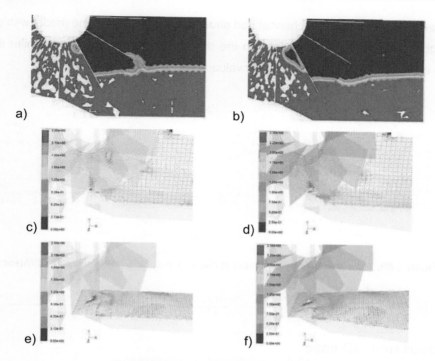

Figure 6-87: Downstream flow conditions

The total pressure monitored at the three points below the wheel hub show variations in both the cases modelled. The pressures for the model with parallel walls lie between -3000 Pa and 9000 Pa. The model with the diffusor shows lower pressures for maximum and minimum values (Figure 6-88). During the passage of the blades, pressure drops to 0 Pa.

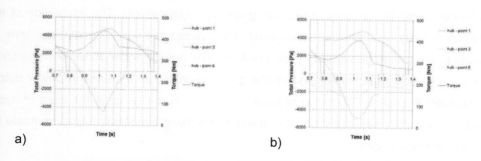

Figure 6-88: Total pressure below wheel hub, a) parallel walls, b) diffusor

The pressure distribution at the channel bed show that the values for the model with parallel walls (Figure 6-89 a)) and the model with the diffusor (Figure 6-89 b)) are similar through the cycle except for the regions where the values drop to their minimum.

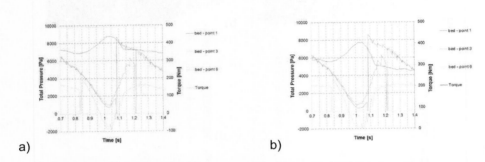

a)   b)

Figure 6-89: Total pressure below wheel at channel bed, (a) parallel walls, (b) diffusor

6.5 Discussion: 3D case

The increase in torque due to the change in blades geometry are for a given discharge and for given water levels in the channel at the upstream and downstream sides of the rotor. For the VOF model, these boundary conditions had to be specified for the runs. As the theoretical values for torque are based on the head difference, the simulations showed that for given head differences, alterations in the blade angle geometry give a better overall performance of the machine. Although upstream water level measurements in the flume were made in a region not included in the computational domain, results are acceptable as the validated show the tendency of torque generation qualitatively. The interaction of the factors flow velocity, upstream water level, rotational speed and downstream water level all contribute to the torque generated by the machine. The results of the 3D simulations show that under given operating conditions (in the cases chosen), an increase in power output can be gained through an alteration of blade geometry. In reality, as in the case of hydro turbines, an individual design for the machine under working conditions has to be made for a specified operating point.

Wall gap

In the 3D model, the extension of the computational domain shows that the wall gaps play an important role in the filling and emptying of the chambers of the rotor. The movement of air enables partially submerged chambers to begin to fill from the gaps without the leading edge of the blades having to enter the free surface. This reduces the height of the upstream wave caused by blade impact. The wall gaps also enable a better emptying of the chambers downstream where air begins to enter the chambers at the hub of the machine, irrespective of the water level in the previous chamber. This results in a smoother curve for torque. However, hydraulic losses incurred through the gaps need to be taken into consideration in view of the geometric tolerances required during the manufacturing process.

Blade angle

For the given number of blades modelled for the runner the alteration of the angle of the blades is a dominant factor. The simulations show that for increasing blade angles the leading edges of the blades enable an improved entry of the blades into the free surface upstream. Recirculation spreads downward from the water surface as the wheel rotates. The entry of the blades is gradual as opposed to the straight blades where the full width of the wheel enters the water surface at the same time causing wave build-up in the flow. The cases with variation in the blade angle show an increase in efficiency with an increase in the angle of the blades to the axis of rotation. The simulation of the angled blade model shows that the geometry has to be adapted to suit the prevalent flow conditions. The cases studied show smoother characteristics for the curve for torque as they are physically more realistic.

Wave build-up

As seen in the results of the simulation with straight blades, the entry of the blades into the upstream surface causes a wave to build up upstream of the blade. The wave propagates upstream due to the impact of the leading edge of the blade on the water surface. For the wheel with gaps at the chamber walls, the wave is higher in mid channel as water is able to flow downstream through the wall gaps beside the wheel.

Recirculation

The model with blades across the full width of the channel shows recirculation in the flow upstream of the wheel. As the flow passes through the funnel-shaped inlet duct there is an increase in the velocity. On reaching the channel, the step in the channel bed causes recirculation. The recirculation in the flow occurs mainly in the vertical plane and follows a cylindrical pattern. The results for the wheel with angled blades show an asymmetry in the zone of recirculation at the channel bed. Additionally, recirculation occurs along one wall of the channel upstream of the wheel. In this region, the recirculation occurs mainly in the horizontal plane. The entry of the blades in the free surface drives the flow towards the walls of the channel. Through the influence of the angled blades the cylindrical recirculation pattern at the step in the channel bed assumes the shape of a cone.

Wheel/channel width

The blade width has a direct influence on the recirculation in the flow due to the asymmetry of the blade geometry. The variation of wheel width alters the recirculation in the flow. For the wheel width being at the maximum channel width, recirculation occurs at the upstream side of the machine. With a reduction of the wheel width there is an increase in efficiency. Recirculation at the inflow into the chambers is reduced which allow for vortices and to pass the machine. Further reduction in the width ratio may lead to higher efficiencies with a reduction in overall power generation. The effect of plates adjacent to the wheel shows a marginal change in efficiency and can be considered negligible.

Bed slope

In the runs with straight and angled blades, the large zone of recirculation at the entry into the channel is caused by the step in the channel bed. Contours and path-lines of velocity show that the step at the inlet of the channel accounts for vortices and turbulence, thereby having influencing torque generation. The introduction of sloped beds for the upstream step and the step directly below the wheel showed has an influence on the performance of the machine. On the upstream side, the recirculation could be reduced by eliminating the step. The remaining recirculation is caused by the angled blades directing the flow towards one wall.

The 3D runs show that flow conditions can be accurately modelled. The models show that for high discharges the effects of three-dimensionality can be represented sufficiently well to enable predictions for performance of machines running under different flow conditions. The modelling of the wall gaps enable a better filling and emptying of the chambers showing lateral variations in the flow. Verification of the machine in the experimental flume has been done based on the given set of boundary conditions. The simulation of the angled blade model shows that the geometry has to be adapted to suit the prevalent flow conditions. Some cases show smoother characteristics for the curve for torque as they are physically more realistic.

For the cases chosen, comparison of the measurement results with those of the simulations indicate that torque generation is dependent on the definition of blade geometry for a favourable working point. Vice versa, given working points may be defined to determine geometrical adaptation of the blades to increase the efficiency of the machine. The simulation models show the capability of predicting efficiency accurately. Based on the flow situation in the channel, an alteration in the channel geometry (walls, bed, etc.) can influence the flow through the machine and an additional increase in power generated may be expected. As the areas of application of the machine are limited to have a minimal environmental impact, alterations in channel dimensions may not be feasible from this aspect. The simulations have shown that a large computational domain and the associated increase in run time are required. The full model was run on a high performance cluster with 16 cores with run times in the order of 36 hours for 10 seconds simulation time. The simulations have also shown that the machine modelled here is robust, with sensitive calculations not leading to a significant increase in performance. Hence CFD modelling for machines like water wheels with variations in geometry do not bring additional benefits. The machines are resilient having efficiencies around 65%-70%.

In conclusion, the validated model shows the capability to predict power generation sufficiently for machine design. The application of CFD provides a visualisation of the flow in detail. For complex turbomachinery CFD techniques provide solutions with a high resolution. Varying mesh densities enable the solutions for high velocity flows. Turbomachinery design can therefore offer performance predictions for machine design.

The simulations performed show that the efficiency of the hydrostatic pressure machine lies at around 68%. The value is relatively low compared to hydraulic turbines. In essence, the variations in channel bed, machine-width etc. has an influence on the flow pattern but the performance is not enhanced. This indicates that the machine is robust in its performance.

Chapter 7
Discussion

7.1 Small hydropower machines

The objectives of the project were to investigate the machines with physical and theoretical models. The three fields of application of the machines were to be developed with 2D/3D numerical models as well as physical modelling tests. The numerical models were to be used to optimise the geometry. Ecological compatibility was to be ensured, conform to the respective guidelines, by the implementation of optimal design in the large-scale models. For developing countries, the use of appropriate technologies was to be considered during the design stage. The turbines developed in the project were suitable for three areas of operation. One area was the deployment in weirs, channels and rivers. The second application was designed for free streams applications while the third field was deployed in water supply networks with low heads. For the installation of the turbine in developing countries, the machine was to be modified suitably. These items are addressed in this chapter.

7.1.1 Numerical and experimental investigations

In this research, CFD simulation software was used to model rotating hydropower machines with free surfaces, using Volume-of-Fluid (VoF) methods that enabled obtaining a better understanding of the physical processes involved. The numerical simulations in Chapter 5 showed that the 2D water wheel simulations are valid for low flow situations. The filling and emptying of the chambers of the wheel in reality takes place in three dimensions, with air entering the chambers as the blade rises off the free water surface as well as air entering the chambers through the wall gaps. The limitations of 2 Dimensional modelling do not allow for these processes to be sufficiently resolved. The 2D models however provided an assessment of required mesh refinement, boundary conditions, solver settings and solution methods to be used in more extensive 3D simulations.

In Chapter 6, the 3D runs showed that flow conditions can be accurately modelled. The results showed that for high discharges the effects of three-dimensionality can be represented sufficiently well to enable predictions on performance of machines running under different flow conditions. The modelling of the wall gaps enable a better filling and emptying of the chambers showing lateral variations in the flow. Verification by experiments was carried out based on a given set of boundary conditions. The simulation of the angled blade model shows that the geometry has to be adapted to suit the prevalent flow conditions. Some cases exhibit smoother characteristics of the curve for torque which is physically more realistic. The simulations showed that a large computational domain and associated increase in run time are required. The full model was run on a high performance cluster with 16 cores with run times in the order of 36 hours for 10 seconds simulation time. The simulations also showed that the HPM model is robust, sensitive calculations not leading to a significant increase in performance. It was observed from CFD modelling that variations in geometry for machines like water wheels do not bring additional benefits. Apparently these machines are resilient to shape variations, all having efficiencies between 65% - 70%.

The numerical model of the machine generated using CFD software was validated with experimental data from the HYLOW project, confirming the capability of predicting performances of hydropower machines by numerical simulation. The results of the simulations show a good match with the experiments. Variations in model geometry can be explored at an early design stage. The performance of the hydrostatic pressure machine under different flow conditions can be determined and adjustments in geometry can be made to achieve acceptable performance. For given flow conditions at a potential site, the machine design can be specified using CFD. No one-size-fits-all solution for hydropower machines is possible and CFD can provide structural design considerations adjusted to the specific hydropower regime. The objective of the HYLOW project was to develop turbines for low head differences below 2.5 m. These small hydropower machines with ratings <1000 kW were considered to cost less than conventional machines.

Machines which utilise the kinetic energy in rivers are found to have efficiencies around 30%. Low flow rates with velocities ranging less than 1.6 m/s hinder good energy harvests. By introducing head differences (pressure difference), the efficiency can be improved. River continuity is ensured by the low rotational speeds of the hydrostatic pressure machine and the large chambers of the machine. Environmental impact can be considered low.

7.1.2 Conversion efficiencies

The development of converter technologies had the objective to develop a theoretical background for design and a prediction of performance of low head hydrostatic converters in rivers. Model testing was done by the HYLOW project partners to verify the basics of various machines. General model testing on various scales were performed to assess scale effects. The hydrostatic pressure machine geometry was adapted to look higher economic and environmental efficiency. River tests with a scale model were carried out to assess performance in terms of fish passage and sediment transport. The installation was monitored and the optimised design was validated. The hydrostatic pressure machine and the simpler hydrostatic pressure wheel were explored by varying the hydrostatic pressure difference between the upstream and downstream water levels. While the hydrostatic pressure wheel consists of a simple wheel with blades emanating radially from the shaft, the hydrostatic pressure machine has a cylindrical hub to which the blades are attached radially. The theory was expanded to include losses due to turbulence and leakage. The hydrostatic pressure machine could be installed for heads in the range of 1.0 to 2.5 m, the smaller wheel for heads of 0.2 to 1.0 m.

Tests performed on three wheels by the HYLOW partners at the Technische Universität Darmstadt and the University of Southampton confirmed the determined theory. As stated in the Project Final Report, initial tests with a small scale HPM at the University of Southampton showed quite promising mechanical efficiencies of up to 81 %. Since the scale was quite small, the HPM was then built on a larger scale (D = 1200 mm) at Technische Universität Darmstadt and tested in a 1.0 m wide channel to assess the effect of different blade geometries and of different wheel width to channel ratios under varying flow conditions and water levels. The model had a hub diameter of 0.40 m and blade depth of 0.40 m. Straight, curved and diagonally fixed blades were installed and tested. A HPM configuration with straight blades mounted diagonally with an angle of 20 degrees on the hub was found to give the best performance. This installation has the advantage that the blades dip relatively steep into the upstream water surface and therefore cause the smallest turbulences on the upstream side. In addition, water lifting on the downstream side of the HPM was prevented. A ratio of 1:2 between the HPM and the channel width was found to be ideal. This constellation allows water to fill the cell volumes not only frontally but also laterally. In this way higher maximum flow rates can be achieved and higher efficiencies can be obtained (Mueller and Batten, 2012a). These aspects were investigated from CFD simulations in Chapters 5 and 6.

The theoretical values were comparable to those obtained from physical model tests. For low flow rates, losses due to leakage influenced performance considerably, as compared to high discharges. It was found that for higher rotational speeds turbulent losses have a large influence on the total losses. Variations of the downstream water levels and the shape of the downstream channel were also investigated. In the model tests it was found that the upstream water level had an insignificant influence on the overall efficiency as compared to the downstream water level. In field tests monitored over a period of 10 months the above mentioned effect was not observed, possibly due to scale effects. The field tests were performed on an installation in an existing mill race in Germany having a head difference of 1.2 m. The machine with a width of 0.82 m and diameter of 2.45 m ran at 10 rpm on a discharge of 0.645 m³/s. The installation underwent the regulatory planning process to obtain the permission to construct the prototype. The environmental impact was assessed and aspects like flow continuity and residual flow rates were addressed in order to comply with the regulations. Electricity was fed into the grid through a rectifier at the generator. Measurements of electric voltage and current were recorded providing the information for calculating the mechanical power output. Losses due to leakage were reduced by the addition of rubber flaps along the edges of the blades.

The installation showed mechanical efficiencies between 0.7 and 0.9 for varying discharges and downstream water levels. An increase in performance was observed with higher downstream water levels similar to those in the model tests. For the installation, wheel rotational speeds above 7 rpm led to turbulence and wave formation on the entry of water into the chambers. This phenomenon could not be observed during the model tests. The installation also showed that a drop in the water level upstream led to an increase in efficiency. This effect too, was not visible during the model tests. As mentioned by (Mueller and Batten, 2012a), the introduction of rubber seals at the edges of the blades reduced losses considerably. In the flume at TU Darmstadt tests were conducted to observe the behaviour of fish in the vicinity of the wheel. In addition, the movement of debris and sediment of different sizes through the 1.2 m diameter wheel was observed. While particles within the area of influence of the blades on the upstream side were drawn into the chambers and exited into the downstream channel, other particles circulated in the upstream zone and were deposited in regions of very low or no velocity. The wheel, while in operation, also allowed floating debris to pass.

The field tests showed that for optimal geometries, efficiencies up to 89% could be reached with 10 to 12 diagonally mounted blades. The ideal blade angle was 20° with the channel width to wheel width ratio of 2:1. The reduction of the gaps between wheel and channel sides as well as bed using rubber strips improved performance and reduced the risk of blockage by particles or debris. The upstream water level is to be kept below the hub to enable better blade entry whereas the downstream water level is to be maintained up to the hub. Results showed that efficiency increased with a larger blade angle. Results also showed that the ratio of the channel width to wheel width has an influence on the efficiency. Recirculation in the flow can be reduced with a smaller wheel width. With asymmetric blade geometry, a narrower wheel allows vortices and other losses in the flow between wheel and channel, thereby reducing recirculation at the inflow into the wheel. As a result, performance is increased. The plates at the sides of the wheel do not have a significant influence on the performance as also seen in the experimental setup at the University of Southampton. The simulations showed that end plates have a marginal effect on the wheel (η = 0.688 compared to η = 0.681 respectively).

A low-cost economic solution for the power take-off enhanced with the possibility of lifting water for irrigation or storage for power generation could strengthen the impact of the machine. The results also showed that there was potential for the implementation of the free stream energy with power ratings from 1 to 3 kW in large rivers. These machines could be implemented in rivers to produce energy for local consumption.

7.1.3 Full scale model experiments

The installation of a full size model provided results of performance in situ. The effects of the machine on the environment were analysed. The design flow and corresponding water levels together with the monitoring of performance and environmental impact were compared with small scale experimental values. Flood risk analysis at the site for the particular design, the permission process, fish passage and sediment transport were all monitored by the partners in the HYLOW project at the Universities of Darmstadt and Sofia, Bulgaria. A machine with a diameter of 2.4 m was installed in a weir in the river Iskar in Bulgaria after obtaining the necessary permissions from the concerned authorities. The rotor was constructed with 10 blades having a width of 2.0 m. The design was based on the small scale machine explored at the Universities of Southampton and Darmstadt. A weir was reconstructed to provide the required flow conditions for operation of the machine, enabling the investigations at full scale. Figure 7-1 shows the machine in situ at the site in Bulgaria (Mueller and Batten, 2012b).

(a)   (b)

Figure 7-1: View of the site in Bulgaria from a) upstream, b) downstream Bulgaria

A generator with a rotational speed of 100 rpm was driven by the runner. A gearbox between the machine and the generator enabled an increase in rpm. For the given rotational speed, the permanent magnet generator had a power rating of 20 kW. The gearing consisted of 2 stages. Torque measurements were taken between gearbox and generator. Electricity generated was fed into a load bank by the stand-alone unit. High performance electronics converter enabled variations in load current ranging from 0 to 48 A in the form of direct current. Measurements of parameters like rpm enabled assessment of performance under varying conditions, thereby giving the optimal operational point. Results showed that for maximum power output the upstream water level needed to be above the hub. With the upstream water level slightly below the hub a maximum efficiency could be obtained. Mechanical efficiencies in the range of 55 to 65% could be obtained at rotational speeds between 7.5 and 10 rpm. Power output between 6 and 7.5 kW were obtained for varying water levels. The hydrostatic pressure machine prototype installation was in the head range of 1 to 2 m. The setup of the prototype provided an alternative to conventional technology. The implementation of the prototype included the complete process of planning a hydropower station, from site location to power generation. It enabled an assessment of the machine in including its environmental impact.

7.2 Other HYLOW project findings

The HYLOW project consortium consisted of several project partners each of whom addressed specific topics which covered various issues connected to very low head hydropower and its influence on the environment. For the development of the turbines theoretical and physical modelling was undertaken. A large-scale test setup enabled design, construction and monitoring of the machines with subsequent decommissioning. In compliance with regulations (Water Framework Directive), the environmental impacts were assessed in terms of biological impact, effects on fish population and sediment transport. Furthermore, the use of appropriate technologies, transfer of technology and the outreach were investigated. Detailed findings are reported in the HYLOW End of Project Publishable Summary (Mueller and Batten, 2012a).

7.2.1 Environmental considerations

Although hydropower is often considered a so called green energy or renewable energy sources, the installations negative impacts on the environment are also observed. Impacts of hydropower on the environment are typically (i) the disruption of river flow continuity; (ii) the reduction of water flow leading to an unnatural flow regime; (iii) the mechanical damage to aquatic life; (iv) the effects on natural sediment transportation. Directives like the EU Water Framework Directive call for measures to counterbalance these impacts on the environment. Ideally, measures need to be taken to enable:

- Maintaining natural fluctuations of flows upstream and downstream for biodiversity
- Remedy migration of aquatic flora and fauna
- Almost natural condition of the river bed and banks
- Assuring continuation of sediment transport
- Taking mitigation measures like installing fine trash racks and fish passes

These issues have been addressed within the HYLOW project. Details may be found in the Published Summary (Mueller and Batten, 2012a). Impacts on the environment were assessed in field tests and details about regulations for an installation were collected from local authorities in Germany (as an example). From the point of view of small hydropower generation in a global context, the EU regulations may not be directly applicable to the rest of the world, but can provide indications how the various technologies can be suited to match local regulations in target countries.

The machines explored within the HYLOW project were intended to be installed in the Europe Union in compliance with EU Directives. The unused hydropower potential in European rivers was estimated to lie at 5 GW, the potential for heads below 2.5 m in Germany at 500 MW and in the UK around 600 – 1000MW. This hydropower potential mainly found at weirs. With the implementation of local norms, this technology cannot be resourced with conventional turbine technology as considerable structural requirements are necessary and the ecological impact is high. Further, conventional turbines like Kaplan turbines at low head sites require large discharges which in turn lead to bigger machine diameters. This results in a greater ecological impact in terms of e.g. endangering fish.

The use of conventional turbines increases the risk of fish mortality and hinders sediment transport, adding a further negative impact on the environment. The EU-WFD prevents the installation of hydropower machines in such cases, thereby making the utilisation of very low head hydropower problematic. Although the Water Framework Directives is mandatory for the EU, the basic concept may be adapted for different countries.

Fish Passages

In the European Union, the Water Framework Directive provides guidelines for the ecological improvement of water bodies. Hydropower sites require a head difference which is usually generated by cross structures like weirs. These cross structures enable the machines to utilise the potential for energy generation, at the same time breaking the flow of the river. This leads to the collection of sediment on the upstream side and scouring on the downstream side. Changes in the ecology are thus introduced with fish migration being impeded. The directives for the preservation of biodiversity and habitats have to be adhered to for renewable energy production. The EU-WFD states that for EU water bodies requirements need to be fulfilled that these water bodies have either a "good ecological status" or, if modified in a large way, "good ecological potential". A reduction in ecological quality is not permitted and any changes should assure the protection, enhancement or restoration of the water body.

The EU Eel Regulations (1100/2007) stipulate that a minimum of 40% of adult eels return to the sea to spawn in annual cycles. Measures should be taken by member states to enable this, which in turn requires appropriate hydropower generation technologies. Other regulations which were considered in the HYLOW project are the Convention on Biological Diversity and the EU Habitats Directive (92/43/EEC). The objective of exploring fish passages was to assess the impact of energy converters in a natural environment by HYLOW project partners. The acoustic impact on fish was investigated in a laboratory flume. The results of the investigations at the installation site were included in the HYLOW Environmental Impact Statement. At the University of Southampton laboratory, the behaviour of brown trout was studied in a model of the free stream energy converter. At the site on the River Warnow in Germany, the investigations were conducted by trapping the fish and using radio telemetry. With the model in the flume, the fish showed a low level avoidance as a response to the hydropower machine. This response did not show significant delay in fish migration when compared to the flume without the model. After the installation, 65% of fish passed the converter in 1.5 hours. Three fish of a total of 186 which passed through the converter had blade strike related injuries. It was observed that the machine did not hinder the movement of fish in both the upstream and downstream directions in the laboratory.

A blade strike model was developed to give an insight into the impact of the hydropower machines, with the option of taking remedial measures. As the machines developed in the HYLOW project have low rotational speeds and do not operate with strong negative pressures compared to conventional machines, blade strike was expected to be low. The hydrostatic pressure converter in the laboratory flume showed that 28% of euthanised brown trout had signs of injury. A total of 48% of the fish made contact with the blades. As the machine was a scale model, blade strike is expected to be higher in the full scale prototype. Fish that were introduced in the laboratory flume were of the species grayling, rainbow trout and European eel. The fish moved downstream towards the wheel. Of the fish, 85% was prevented from entering the wheel by a screen. In an external flume eel, chub and other fish were introduced upstream of the machine. Here too, the fish moved downstream towards the wheel. In both cases the acoustic environment created by the operating wheel did not seem to deter the fish from entering the wheel. Hence screens were needed to be incorporated in the design to divert fish to the fish passages.

The movement of fish in the Iskar River in Bulgaria was explored by netting and using Passive Integrated Transponder telemetry. Fish mortality was found to lie around 2% for fish up to 220 mm. For larger fish, the mortality rate was found to lie at 26%. For euthanised fish, the strike rate was found to be around 9%. The higher rate for live fish was due to their behaviour. Fish moving downstream face the current while doing so. They move at a lower velocity than the flow, thereby increasing the possibility of blade strike. With the machine in operation, half the fish moved downstream using the fish passage.

With the installation of the hydrostatic pressure machine, river continuity is interrupted. This can cause delay in fish movement up and downstream. In this case fish passages or ramps which are placed in regions of high velocity like tail races can provide free movement of fish. Apt screens to prevent fish from entering the machine need to be installed to protect long bodied fish like eels from being damaged. It was found that, for the free stream energy converter, the gaps in the machine and the draught below the machine also contribute to the reduction in damage to fish. This machine also has the advantage of not having to break river continuity in order to generate the head difference. It was seen that a gap between the blades and the structure reduced the probability of blade strike.

Morphodynamics

The geometry of the hydrostatic pressure machine allows for the passage of sediment (bed load and suspended load) through the channel and for deposition downstream. Observations on sediment transport and deposition were done with a physical model in order to provide recommendations for the installation as a run-of-river installation by the HYLOW project partner at the University of Braunschweig. Physical models were used to explore ways of minimising the effects of trapped sediment transport and deposition downstream. The morphodynamic regime near small run-of-river installations was investigated in generic model tests. Different morphological boundary conditions were considered. Scouring at the inflow wing wall of the HPM installation and the redirection of sediment around the HPM installation and over the weir were observed. Figure 7-2 (Mueller and Batten, 2012b) shows the influence of the inlet structure on the upstream side.

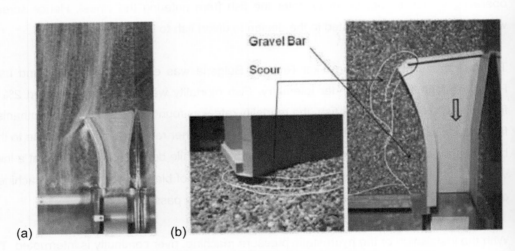

(a)   (b)

Figure 7-2: Upstream effects in the model (a) streamlines (b) effects on bed

During the investigations at the weir, scouring was observed upstream of the weir and on the upstream side of the wall of the installation. A further effect observed was scouring and the movement of sediment around the installation and transport of sediment over the weir (Figure 7-3) (Mueller and Batten, 2012b).

Figure 7-3: Scouring in the upstream bed

7.2.2 Free stream energy converter

The free stream energy converter developed by the HYLOW partner University of Rostock, consisted of the design and construction of a large-scale model of the energy converter. Local sites for the deployment of the prototype and the choice of testing equipment for measurements during the monitoring of the machine provided data for the validation of a numerical model. Aims of the deployment of the prototype included fulfilling geometry requirements for mooring, stability, and dynamic positioning of the floating converter. Design for power take-off as well as transportation and the deployment of the machine to the chosen sites were also addressed.

Various tests were performed on small scale models, initially in a wind tunnel and subsequently in a small flume. The position of the paddle wheel, gap widths between wheel and bottom plate, relationship to the draught and other variables were tested in three variations of the models in four different flumes and tanks. A model of medium scale was constructed to validate the small scale model results. The medium scale model was tested in a towing tank. The position of the wheel in relation to the length of the model, floating stability, influence of flow velocity on the buoyancy, stability and trim and the influence of flow separators at the model stern were determined. The small and medium scale models provided the detailed dimensions for the construction of a prototype of 7.6 m length and 2.4 m width. As in the case of the installation of the hydrostatic pressure machine, two sites were identified for the machine at the Naval Base in Warnemunde and at the Warnow River, both in Northern Germany. Towing tests were performed at the former site while the machine was moored at the latter. The following Figure 7-4 and Figure 7-5 (Mueller and Batten, 2012b) show the free stream energy converter as a CAD model and as a prototype.

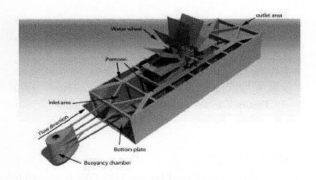

Figure 7-4: CAD model of the free stream energy converter

Figure 7-5: Machine prototype

During the towing tests, a maximum power output of 600W was measured for a towing velocity of 1.5 m/s. The highest power output and efficiencies were measured at rotational speeds between 3 and 5 rpm. The tests showed that an additional element was required at the bow of the prototype to maintain buoyancy. This element was incorporated in the model for the tests in the River Warnow. In these tests with flow velocities at around 1 m/s, a maximum power output of 350 W was measured under near-natural conditions. Rotational speeds ranged between 3 and 4 rpm. Peak efficiency was found to lie at 30% relative to the total width of the machine.

The floating device can operate in rivers and can generate electricity with an efficiency of 40%. Operational flow velocities are between 1.5 and 1.8 m/s. In navigable rivers, installations will be difficult to implement. For large rivers, the potential for local power generation is high. In ecological terms, the free stream converter has a minimal impact thereby making it an attractive renewable energy source. Flow velocities recorded in the river around the machine showed no significant changes indicating a low environmental impact. Changes in the bed level directly below the machine showed an accumulation of sediment of around 20 cm. As stated in the HYLOW Publishable Summary (Mueller and Batten, 2012a), the river bed changes are also based on the natural seasonal dynamics.

The model of the free stream energy converter was set up to study the behaviour of the machine operating in rivers. The numerical model was verified with the results of the experiments. Variations of the model were simulated with different number of blades with varying thickness. The simulations of the free stream energy converter showed the influence of the internal draft on the efficiency and showed that optimum design is directly influenced by blockage in the flow. The bed roughness is another factor of influence which requires design optimisation of the machine for different river installation. Following up from the results of the hydrostatic pressure machine and the free stream energy converter, a simplified power take-off machine was developed (Figure 2-1).

Figure 7-6: Hydrostatic pressure wheel in the laboratory flume

This simplified hydrostatic pressure wheel was found to have potential in developing countries at small sites. Such machines could operate over a large range of rotational speeds with efficiencies of up to 90%. The construction is simple, enabling local manufacturing. The prototype of the hydrostatic pressure wheel showed that the machine could be installed in channels for head differences between 0.2 and 1.0 m for power supply to households or small settlements.

7.2.3 Micro-turbines in water pipe networks

The third type of energy converter explored in the HYLOW project was the development of turbines inside water distribution systems by the partner at the Technical University of Lisbon. The concept of the turbine was to be realised and installed in water distribution systems. Model tests and CFD analyses were provided the final geometry. A series of tests were conducted to determine the final geometry, design and application range together with performance curves.

A series of tests was made with a tubular propeller turbine in an existing water supply system in the town of Mafra in Portugal after obtaining the necessary authorisation (Figure

7-7) (Mueller and Batten, 2012b). Laboratory tests generated 65 W of power for a discharge of 12 l/s and a head of 1 m. The tests in the distribution system were performed over a 24 hour period to study the hydrodynamic behaviour of the system for varying flow and discharge conditions. The results showed the performance of the turbines over a cycle of one day. The positive displacement turbine was tested in the laboratory and showed that applications had low rotational speeds and high torque, thereby requiring gearing to increase the speed to the generator.

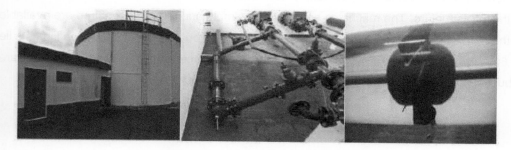

Figure 7-7: Water supply network in Portugal

Energy generation using a pump-as-turbine was investigated in the laboratory and in a water supply system (Figure 7-8) in Portugal for two scenarios. In the first case of steady flow, power generation increased with discharge, the measurements being close to the theoretical values. In the second scenario under runaway conditions, pressure surges at the same flow rate increased the rotational speed. The results showed that the system operates at an efficiency of 50% for high pressure flows.

Figure 7-8: System tests in pipes

The energy converters for such closed water supply systems could be constructed using commercial equipment. The potential in water supply systems could be exploited 24 hours per day, thereby contributing to emission-free power generation. There is no impact on the environment, the system being closed. A study of the potential of the pump-as-turbine in water supply systems in Portugal showed that installations were economically feasible at numerous sites, with some operators showing interest in the potential of generating renewable energy. The converters could provide local energy solutions to networks not connected to national grids. For developing countries these systems could provide a source of energy and contribute to local energy supply. The micro turbines can be used to generate power in existing water supply networks. They do not require significant changes in the infrastructure. The machines are relatively simple and the parts required are available in the market. The return on investment period is below four years and reduction in costs lie at 60% for the lifespan of the micro turbines. Maintenance is minimal as the turbines are robust and electricity can be produced round the clock. (Mueller and Batten, 2012a)

7.3 Small hydropower in a global context

The use of low head hydropower can contribute to a reduction in use of fossil fuels. Islands can be suitable niche markets for providing energy independence. The exploitation of low head hydropower can be implemented in a local setting with the consumer not having to have long transmission lines. Power production lies in the range of 50 to 70 kW. With only a part of the very low head hydropower resources being explored in Europe, the scope for small hydropower seems high.   The main reason for this is because conventional technology does not fulfill the ecological requirements. The technology is not economically feasible. Hence the HYLOW project aimed at developing new technologies to for hydropower generation using very low head differences.

7.3.1 The European perspective

The EU Water Framework Directive stipulates that water bodies and water courses cannot be altered in a negative manner in terms of ecological impact. Various aspects of the HYLOW hydropower converters were considered with respect to the directive. Modifications of the machine as well as the effects of the directive on the development of low head hydropower were also considered. The technical evaluation of the machines by relevant authorities and an analysis of the free stream potential in coastal and estuarine regions in selected countries were addressed by the HYLOW partner University of Rostock. In general, hydropower installations have a negative ecological impact on rivers and river systems. The break in the continuity of the river or stream prevents the free movement of sediment downstream. This is also influenced by the altered flow regime in the system. Mechanical damage to the flora and fauna occurs within the turbines. The water framework directive requires that these above mentioned negative effects of hydropower plants are minimised in order that the water body maintains a good status or potential. Comparison of the state of the water body with that of a reference enables a categorisation of the status. The morphological, biological and chemical composition is also considered by limiting for e.g. emission values. Hence a low impact of the installation on the environment improves the status of the river. The directives objectives are to ensure that the river bed and banks are kept in a natural state with sediment continuity. Upstream and downstream fluctuations in the river water level are to be as close to the natural state as possible. The movement of flora and fauna is to be enabled, even if, in the case of fish, a passage is required. The structure may require modifications to ensure that flow rate, flow velocities are maintained.

The field tests on the hydrostatic pressure converters gave an estimate of the impact of the machines on the environment. The tests enabled a comparison of theoretical and measured values. The tests also provided a basis for discussion with the authorities responsible for permission of the installations. It was found that the impact of the free stream converter had, at most, a minimal negative influence on the environment. Very minor changes were observed due to the mooring of the machine, thereby requiring no changes with regard to the water framework directive. In comparison, it was found that the hydrostatic pressure machine required a fish passage to enable the migration of fish upstream. For small fish and sediment, the machine does not obtrude passage. The risk of injury to the fish is low and medium sized sediments do not damage the machine. To prevent damage to larger fish, the installation of a fish screen is required.

As seen in the prototype during the field tests, the HPM machine has the potential to expand into new areas of power generation. One scenario had been set up for the UK and Germany regarding the economic impact of local low head power generation. Utilising one half of the available potential in these two countries and with costs per kW capacity at €4,500 give a market value of 2.2 billion Euro. For the UK, assuming a feed in tariff of 19.7 pence per kWh, the annual production value lies at 638 million Euro. Considering that the hydrostatic pressure machine is manufactured in small factories, the units could contribute to employment in the developing world. The revenue earned by site owners would provide them additional income. Assuming a similar scenario in other regions of Europe, a potential of about 3 GW may be utilised for power generation. With a rating of 100kW, the total potential could be distributed over 30,000 sites. For installation costs of €4,500 per kW, the market value would lie at 750 billion Euros. (Mueller and Batten, 2012a)

The Publishable Summary also reports a study of potential sites for installations of the hydrostatic pressure machine which showed that information of head differences (0.6 to 2.5 m) was accessible for about 6,440 cross structures in Germany from a total of 100,000. For known discharges in the river Hunte in Lower Saxony, a power potential between 1.7 and 7.5 MW exists at 30 structures. The results of the study of sites in Lower Saxony may be extrapolated to other regions having a similar terrain. For Bulgaria, data showing the heights of cross structures was available for 89 sites with discharges between 0.2 m³/s and 53 m³/s. Mean annual discharges for 55% were less than 3 m³/s. For 55% of the sites, the average annual discharge was 3 m³/s. In the Netherlands, the regional Water Authority "Brabantse Delta" in South Netherlands, identified 9 suitable locations.

On a general note, new technologies for power generation offer capabilities with significantly less environmental impact, simultaneously broadening the possible areas of application. The ability to produce electricity at a decentralised level reduces the dependence on central power generation with improved energy security. Power generated in decentralised installations can be used locally.

7.3.2 The global perspective

Since technologies for exploitation of hydropower for heads less than 2,5m and low flows is not economically feasible for conventional high efficiency turbines like Kaplan, the machines developed in the HYLOW project were considered as an extended area of application of hydropower generation. The machines have an ecological advantage which allows for the utilisation of additional resources. The HYLOW project has shown that the development of the hydropower machine has the potential to provide cost effective solutions for the generation of hydropower utilising very low heads. The installations showed that the technology performed well in real field applications. Decentralised energy production for local consumption reduces the high investment costs involving power distribution thereby providing income for local stakeholders in the region. These relatively simple machines do not require significant installation and maintenance costs. During the course of the project, generated electricity has been fed into the grid. The project showed that the economic and environmental aspects as well as the efficiency of these converters are good. The efficiency can be considered acceptable for these fields of application. Traditional water wheels like the Zuppinger wheel have rotational speeds between 3 and 7.5 rpm and large diameters between 4 and 7.5 m. The low speeds require gearing which increase the costs to between €10,000 to €12,000 per kW of capacity. Fish compatibility information is not available and sediment passage is not possible.

Turbines running at low rotational speeds have significant ecological advantages over conventional technologies. The energy converters developed in the HYLOW project allow for river continuity. The low speeds reduce the potential damage to fish which can pass through the machine. The flow is not accelerated; pressure differences are minimal which do not adversely affect fish. Outflow velocities are not as high as conventional technology making it easier for migrating fish to use the guiding screens to fish passages. The river continuity ensures that sediment can be transported through the installation. As environmental issues play a large role in hydropower installations, these energy converters have a distinct advantage over conventional turbines thereby increasing the area of application. The applications of conventional technology for very low head hydropower generation have long pay-back periods. For low power ratings between 5 kW and 100 kW, unit costs are high with return-of-investment periods ranging between 10 and 25 years thereby making investments barely economical. For these technologies, maintenance costs also increase overall costs. As reported in the Publishable Summary, power demands for households and small settlements lie in the range of 0.1 to 5 kW. The power requirements could be in the form of electrical, hydraulic or mechanical energy. Power demand and distribution is shown in Table 7-1.

Table 7-1: Demand structure and distribution distances (Mueller and Batten, 2012a)

	Electricity [kW]	Mech. Power [kW]	Hydr. Power [kW]	Distribution distance [m]
Single household	0.1-0.5	-	-	< 100
Small settlement	1-5			< 100
Agriculture	-	1 - 2	-	<100
Workshops / businesses	0.2 - 2 (lighting, recharging)	1 - 2	-	< 5 (mech. power) < 100 (El. Power)
Water purification	-	-	1 - 5	< 30 - 50
Irrigation / water supply	-	-	1 - 5	10 - 1000

The simplicity of the construction for power take-off is of importance from maintenance and economic considerations. In this context an assessment of the HYLOW technology showed that the micro turbine power generation systems could not be implemented in developing countries due to the complexity of the systems. Here too, the machines do not have a large market potential. It was also found that the hydrostatic pressure machine was not apt for very low heads. The generation of electricity using the machine was found to be too complicated and not appropriate for developing countries. The installations generating power for local consumption enables energy autarchy. For developing countries, investments in high-tension transmission lines can thus be eliminated so local power generation can improve living conditions in remote settlements.

Sites for very low hydropower installation

In developing countries like Pakistan, large irrigation systems exist with numerous head differences between 0.5m and 3.0 m. The available hydropower with head differences below 3.0 m is estimated at 19 GW. Based on available data for Khyber Patunkhwa Province, 1108 cross structures were available. Based from discharge data, the total power potential at full discharge was estimated to be approximately 28 MW. The sites are situated in six districts of the province. (Mueller and Batten, 2012a) The converters could provide local energy solutions to networks not connected to national grids. For developing countries, the small hydropower systems could contribute to local energy availability. The overall benefit of small installations offer area-wide access to energy. The nexus *energy–availability–health–education–technology–improvement* of living standards can bring along a transformation in society. The socio-economic impact of this access is difficult to measure as it cannot be assessed in numbers. Projects in developing countries have shown considerable success in improving the energy availability as well as overall contribution. The approach has been to consider energy production as the driving force behind societal progress. Appropriate technologies can exploit local resources, provide electricity to first-time users thereby increasing the quality of life significantly. Hence, the use of alternative technologies needs to be considered in terms of sustainability in spite of higher costs. The management and consumption of the produced electricity through cooperatives and local bodies require learning skills which in turn alleviate the communities and provide them an alternative to burning fuels. The overall impact in terms of carbon dioxide reduction in electricity generation is large considering the hydropower potential available. However it can be expected that only a part of the small hydropower potential of can be effectively used for power generation since conversion losses need to be considered. Changes in national government policies as in the case of small hydropower can enable various stakeholders to take part in power generation, thereby ensuring energy availability, creating manufacturing opportunities and reducing fossil fuel consumption.

Chapter 8
Conclusions and recommendations

Rising energy demand for a growing population is one of the challenges in the near future. With sustainable development in the foreground, energy production using renewable sources has gained importance. To meet the energy demand, all forms of renewable energy need to be exploited. The impact of large-scale hydropower projects in monetary as well as environmental terms has opened new avenues for power generation using small hydropower units. Here the adverse effects of large hydropower (silting, deforestation, etc.) are reduced.

Changes in national policies are enabling small hydropower generation in many countries to cover growing demand. The energy production is meant for local grids and can provide area wide coverage. Variability of very small installations and their low impact on the environment make them attractive as energy sources. Further, the involvement of local stakeholders who generate electricity for local consumption enables decisions to be taken at a local level.

Access to electricity in turn brings numerous benefits in different forms such as lighting and the pumping of water. The availability of these resources enables an overall improvement of living standards and enables development in terms of fulfilling the Sustainability Development Goals.

In the small hydropower sector, electricity generation using very low hydraulic heads has not been exploited due to the lack of appropriate technology. In the HYLOW project, three machines were developed and tested using physical and numerical models. The numerical models of the machines generated using CFD software could be simulated and validated with experimental data from the HYLOW project. The results show the processes occurring in the flow during the operation of the wheel. They also enable a visualisation of flow patterns and a better understanding of flow conditions during operation.

8.1 Research Answers

The role of small hydropower in a global perspective is the ability to provide local energy solutions.

The increase in energy demand and the awareness of sustainability has led to the development of clean energy systems. Of these renewable energy sources, only hydropower produces constant energy at all times barring extreme events like drought or floods. Amongst the renewables, hydropower machines have the highest efficiencies. Large hydropower plants are expensive in financial terms as well as in terms of their impact on the environment. These plants require extensive hydraulic structures like dams and need large-scale grids to transport electricity over long distances. Conventional low head hydropower is also relatively expensive as the turbines have to be designed and manufactured according to individual specifications. The effects on the environment are not as adverse as of large hydropower. The potential for energy harvest from hydropower has not been fully exploited, especially in developing countries. Hydropower generation using small machines comes as an alternative as the investment costs in sum are considerably lower than large turbines. They can run off the grid and can widen the spectrum of hydropower generation. The devices used to generate electricity can be installed on a local scale to cover local demand thereby eliminating the need for transportation infrastructure. Programs like the SDGs on an international level as well as other efforts on national and regional levels can promote local hydropower generation thereby improving the overall welfare of the inhabitants of the region. In summary, hydropower can contribute to the growing power demand by providing local energy availability.

2D models enable quantitative analyses of symmetrical hydropower machines for small channel-width to wheel-width ratios.

The 2D analysis of the flow conditions show the processes occurring in one plane of the domain. This simplification enables the visualisation of the processes occurring during the operation of the wheel. The 2D models provide estimates for model parameters like mesh density, solver settings and solution methods to be applied for the solutions. The simulations can explain phenomena like the fluctuations in torque during the course of the runs. The 2D analyses simplify the processes that occur in reality by providing information in one plane. For changes in the width of the channel, transverse flows or asymmetric geometries 2D models are not sufficient to represent the cases. Here 3D analyses are required.

The parameters that dominate the hydropower take-off capability are the upstream and downstream water levels and the discharge, with the rotational speed being a function of the discharge.

In 3D, flow conditions can be accurately modelled. The models show that high discharges can be represented sufficiently well to enable predictions for performance of machines running under different flow conditions. The filling and emptying of the chambers are determined by the geometry of the wheel which in turn influences the flow on the upstream and downstream sides of the channel. The transient nature of the behaviour of the free surfaces upstream and downstream of the hub combined with that of the emptying chambers show the influence on fluctuations of torque generated. These effects are influenced by the water levels upstream and downstream of the wheel. Gaps between the channel walls and the wheel enable a better filling and emptying of the chambers resulting in a smoother curve for torque at the shaft. The blade angle enhances this effect but leads to recirculation in the flow on the upstream side of the wheel. A reduction of the wheel width for a given channel width and a change in the bed slope reduces the recirculation in the flow. Thus the combination between the wheel and the channel geometry influence the flow on both the upstream and downstream sides of the flow. The rotational speed is determined by the discharge and the geometry of the wheel.

Numerical simulations provide high resolution images of the hydrodynamics: in this way CFD can provide analyses of flows in turbines like water wheels.

With today's demand for clean energy, hydropower turbines as well as several innovative hydropower machines with free surfaces are being simulated using CFD methods. These applications enable modelling of the hydrodynamics of multiple free surfaces as in chambers of rotating wheels. The free surfaces can be modelled with a high resolution. Simulations of the machine modelled here have provided details of the processes occurring in the flow in the form of contours and profiles in both phases of the flow. The simulations have shown that a large computational domain and the associated run time are required. The full model was run on a high performance cluster with 16 cores with run times in the order of 36 hours for 10 seconds simulation time. The simulations have also shown that the machine modelled here is robust, with sensitive calculations not leading to a significant increase in performance. Hence CFD modelling for machines like water wheels with variations in geometry do not bring additional benefits. The machines are resilient having efficiencies around 65% - 70%.

Increasing computational capacity with every new generation of computers enables more detailed CFD modelling of the hydrodynamics of sensitive conventional turbines. CFD can enable the development of turbomachinery, predict performance and supply the dimensions for prototypes. It is concluded that small hydropower using appropriate technologies has its advantages for the societal changes it can bring along. Even though the power generation of small hydropower is low, the versatility of deploying small units like water wheels in a stand-alone mode is high. Substantial adjustment can be made on the units to accommodate ecological factors as well as provide satisfactory technical solutions for on-site implementation.

The efficiency of local power generation is not influenced by transmission losses over long distances. The availability of untapped water resources in developing countries can contribute to carbon-free energy and an overall improvement of communities in remote areas. As a general conclusion, it can be said that the role of small hydropower is gaining importance. Power demand is driving innovation in this field. The development of power generation machinery using CFD software enables to optimise performance of energy generation.

Small hydropower generation using appropriate technologies can bring about social change enabling the community to decide how they can improve their living standards.

Large-scale hydropower is deployed in areas with sufficient infrastructure requirements like advanced control systems and high-voltage transmission grids, the stakeholders being multinational or national power generation companies. Staff requirements are often not met by the local population through their lack of qualification. Small hydropower, on the other hand, can fill this niche in the developing world. Due to size considerations, the impact on the ecological sustainability can be maintained. Depending on the technology applied, hydropower generation without dams allows the resource to be reused downstream. Connecting remote places to an existing grid can often be more expensive than local mini-grid solutions or off-grid solutions. Combined with high-performance electronics, small hydropower generation offers both the possibilities of feeding electricity into a local grid or storing the excessive power in batteries. The units may also be used for pumping water.

For the case of very low heads, machines like waterwheels can provide electricity for local consumption. In remote regions these machines can also operate as grain mills. In regions where crops are grown, hydropower machines can be introduced in irrigation channels to generate electricity. Based on irrigation networks, the machines can be adapted for specific channel dimensions. Depending on the discharge, several units may be installed in parallel. The requirements for the machine from design to construction and assembly need to be defined depending on the machine size and type. Based on the number of parts available locally, or parts which may be imported, a comprehensive list with respect to price can be made. This will form the basis for cost analysis and estimation and will provide for the pay-back period of the machines. Smaller machines have higher specific costs but lower environmental impact.

Economically the cost per kilowatt for generation is relatively high but brings along other benefits which cannot be assigned a monetary value. With adequate environmental measures, small units can integrate energy production with other long-term benefits like health and education. Many large projects face strong public resistance as the resentment against decision makers is large. Often the affected people feel they have been deprived of their voice in issues which personally affect them. Introducing small hydropower machines into channels around which villages lie could be an opportunity to involve the local inhabitants, making them stakeholders in the investment and the running of the power plant. For very low voltage energy production the risks of injury are low. The operation and maintenance of the installations can be executed by the local community after having been given some basic training. Thus the responsibility of monitoring the energy production and distribution lies with the stakeholders. These could be the community itself. This could impart to them a feeling of responsibility. Ideally, this could generate employment and bring about the subsequent social change. After the machine costs have amortised, some revenue may be gained by the stakeholders. The engineering aspect in a social context has far reaching consequences in rural regions. Improvement in the overall conditions in regions that do not have electricity can only achieved through engineering solutions. The nexus *energy availability–health–education–technology–improvement of living standards* brings along a transformation in society. The socio-economic impact of this nexus is difficult to measure as it cannot be assessed in numbers. Projects in developing countries have shown considerable success in improving the energy production as well as overall contribution in several projects. The approach has been to consider energy production as the driving force behind progress. The impact of energy on the daily life of the community for example the access to water, can lead to gender equality as in many societies time-consuming tasks like fetching water or collecting firewood have traditionally been done by females. Energy availability (SDG Goal #7) could give them more time for schooling, lighting could enable them to study after dark. Through the introduction or electrical or mechanical devices to replace manual labour, more time may be available for meaningful work, thereby leading to gender equality (SDG Goal #5).

By achieving Sustainability Goal #7, local small hydropower generation offers a solution to encompass other goals and bring about social change in small communities. Technology used in small hydropower is sustainable and needs to be explored further in spite of higher costs.

8.2 Recommendations

Further work on very low hydropower generation with machines like water wheels and should include the entire design chain. From a hydraulic aspect, specific blade design could reduce the losses on the entry of the blades upstream of the wheel. Guide vanes before the wheel could further influence the filling of the chambers on the upstream side and reduce the amplitude of torque fluctuations at the shaft. A variation of the machine drive shaft to flow direction within the channel combined with the blade profiles could also reduce the entry losses into the chambers on the upstream side of the wheel and enhance its performance. The vorticies formed before the wheel show regions of high velocities. Variations in the geometries of the blades and hub could improve performance. The impact of these variations on the flow regime needs to be considered.

The load distribution on the blades and the structural mechanics of the wheel need to be investigated. Further work on the structural analysis for the wheel and its mounting is required along with the choice of materials for a robust design. Investigations in power take-off and gearing alternatives to reduce fluctuations in the torque at the shaft need to be made. Combinations of belt transmissions with electronic inverters or other gearing solutions need to be investigated to improve overall performance.

A detailed cost report depending on the location of deployment needs to be made. Here, the requirements for necessary civil works in modifications within the channel need to be considered. The availability of machine parts and civil structures together with skilled labour also needs to be investigated. Electrical requirements for the storage of energy need to be taken into account. For implementation in remote regions, the availability of equipment at the closest location needs to be studied. Considerations for the employment of local staff for construction and maintenance work need to be taken into account. Implementation of prototypes in developing countries should be in abidance with local regulations and permissions from the authorities in order for development programs to enable funding for such projects.

References

AHEC (2005) Micro Hydro Quality Standards. Roorkee.

Ansys Fluent (2012) 'Theory Guide 12.0', ANSYS Inc., (April), p. Chapter 17.

Bach, C. (1886) Die Wasserräder. Stuttgart: Konrad Wittwer.

Batchelor, G. K. (1967) An Introduction to Fluid Dynamics. Cambridge University Press. doi: https://doi.org/10.1017/CBO9780511800955.

Bhat, V. I. K. and Prakash, R. (2008) 'Life Cycle Analysis of Run-of River Small Hydro Power Plants in India', The Open Renewable Energy Journal, 1(1), pp. 11–16. doi: 10.2174/1876387100901010011.

Bozhinova, S. and Hecht, V. (2012) 'Hydropower converters with head differences below 2 . 5 m', in Proceedings of Institution of Civil Engineers, pp. 1–13. doi: 10.1680/ener.11.00037.

Cauchy, A. L. (1827) Théorie de la propagation des ondes à la surface d'un fluide pesant d'une profondeur indéfinie. Paris.

Ebrahimian, E. (2003) Community Action to Address Climate Change: case studies linking sustainable energy use with improved livelihoods.

Erdelen, W. (2009) 'Presentation by Martin Lees Secretary General of the Club of Rome to The UNESCO Natural Sciences Sector Retreat " Facing the Connected Challenges of the 21 st Century ."', (June), pp. 1–20.

EU-FP7 (2007) 'Seventh Framework Programme Cooperation Theme 5 - Energy: Call FP7-ENERGY-2007-1-RTD Topic 2.7.1: New or improved hydro components and concepts'.

European Parliament (2009) 'Directive 2009/28/EC of the European Parliament and of the Council of 23 April 2009', Official Journal of the European Union, 140(16), pp. 16–62. doi: 10.3000/17252555.L_2009.140.eng.

Froehlich, J. (2006) Large Eddy Simulation turbulenter Stroemungen. Wiesbaden: B. G. Teubner Verlag.

Gansel, P. P. et al. (2014) 'Influence of Meshing on Flow Simulation in the Wing-Body Junction of Transport Aircraft', in Dillmann, A. et al. (eds) New Results in Numerical and Experimental Fluid Mechanics IX: Contributions to the 18th STAB/DGLR Symposium, Stuttgart, Germany, 2012. Cham: Springer International Publishing, pp. 3–11. doi: 10.1007/978-3-319-03158-3_1.

Giesecke, J; Mosonyi, E. (1997) Wasserkraftanlagen. Berlin Heidelberg: Springer.

Giesecke, J; Mosonyi, E. (2009) Wasserkraftanlagen. 5th edn. Berlin Heidelberg: Springer.

Giesecke, J., Heimerl, S. and Mosonyi, E. (2014) Wasserkraftanlagen (in German). 6th edn.
    Berlin Heidelberg: Springer-Vieweg.

Grambow, M. (2013) Nachhaltige Wasserbewirtschaftung. Edited by M. Grambow.
    Wiesbaden: Springer-Vieweg.

Grote, K.-H.; Feldhusen, J. (2007) Dubbel Taschenbuch fuer den Maschinenbau. 22nd edn.
    Berlin: Springer.

Hennicke, P. (2005) 'Long term scenarios and options for sustainable energy systems and
    for climate protection: A short overview', International Jornal of Environmental Science
    and Technology, 2(2), pp. 181–191.

Hirsch, C. (1990) Numerical Computation of Internal and External Flows, Vol. 1. Chichester,
    UK: John Wiley & Son Ltd.

Hirt, C. W. and Nichols, B. D. (1981) 'Volume of fluid (VOF) method for the dynamics of free
    boundaries', Journal of Computational Physics, 39(1), pp. 201–225. doi: 10.1016/0021-
    9991(81)90145-5.

Huppes, G. (2007) 'Why we need better eco-efficiency analysis', From Technological
    Optimism to Realism. Technikfolgenabscatzung -Theori und Praxis (TATuP), 16(3), pp.
    38–45.

HYLOW (2012) HYLOW. Available at: www.hylow.eu (Accessed: 12 January 2016).

Hyman, J. M. (1984) 'NUMERICAL METHODS FOR TRACKING INTERFACES* James M.
    HYMAN', Interface, 12, pp. 396–407.

Ibrahim, R. A. (2005) Liquid slosh motion. Cambridge: Cambridge University Press.

IRENA (2012) IRENA International Renewable Energy Agency RENEWABLE ENERGY
    TECHNOLOGIES: COST ANALYSIS SERIES.

ITPI (2007) Annual Review. Puducherry, IND.

Juhrig, L. (2013) 'Wasserkraftprojekte', in Heimerl, S. (ed.). Wiesbaden: Springer
    Fachmedien Wiesbaden, pp. 327–332.

Ketlogetswe, C. (2009) 'Lessons and Challenges Encountered in the Implementation of
    Solar Energy – The Case of Botswana', Renewable Energy, (267), pp. 79–83.

Khan, M. J., Iqbal, M. T. and Quaicoe, J. E. (2008) 'River current energy conversion
    systems: Progress, prospects and challenges', Renewable and Sustainable Energy

Reviews, 12(8), pp. 2177–2193. doi: 10.1016/j.rser.2007.04.016.

Kleefsman, K. M. T. and Veldman, A. E. P. (2004) 'An improved volume-of-fluid mehtod for wave impact', European Congress on Computational Methods in Applied Sciences and Engineering, ECCOMAS 2004, p. 19 pp.

Lakehal, D., Meier, M. and Fulgosi, M. (2002) 'Interface tracking towards the direct simulation of heat and mass transfer in multiphase flows', International Journal of Heat and Fluid Flows, (23), pp. 242–257.

Landustrie, B. (2016) Landustrie. Available at: http://landustrie.nl/nl/home.html (Accessed: 15 November 2016).

Lecheler, S. (2009) Numerische Strömungsberechnung. 1st edn. Wiesbaden: Vieweg & Teubner.

Melorose, J., Perroy, R. and Careas, S. (2015) 'No Title No Title', Statewide Agricultural Land Use Baseline 2015, 1. doi: 10.1017/CBO9781107415324.004.

Michalena, E. and Hills, J. M. (2013) 'Renewable energy governance: Complexities and challenges', Lecture Notes in Energy, 23. doi: 10.1007/978-1-4471-5595-9.

Miller, H. and Escher Wyss (1974) 'The STRAFLO Turbine: The realisation of Harza's idea'.

Ministry of Law and Justice (2003) 'THE ELECTRICITY ACT, 2003 [No. 36 OF 2003]', (36).

MJ2 Technologies (2016) VLH Range , MJ2 Technologies, France.

Mohapatra, G. Ali, A. Mukherjee, S. D. (2009) Renewable Energy for Rural Livelihoods in MNRE-UNDP-FRG Project Villages in Rajasthan and Uttarakhand : A Report, Ministry of New and Renewable Energy, Government of India. New Delhi.

Mueller, G. and Batten, W. (2012a) HYLOW End of Project.

Mueller, G. and Batten, W. (2012b) Hylow Publishable Summary - Figures.

Osborne, D Cutter, A Ullah, F. (2015) Universal Sustainable Development Goals - Understanding the Transformational Challenge for Developed Countries.

Ossberger GmbH + Co (2015) Ossberger GmbH + Co.

Pollio, V; Frontinus, S. J. (1528) M. Vitrvvii De architectura libri decem : summa diligentia recogniti atq; excusi : cum nonnullis figuris sub hoc signo * positis, nunq antea impraessis : additis Iulij Frontini De aqueductibus libris, proter materiae affinitatem. Lyon: Guill. Huyon.

Pujol, T. et al. (2010) 'Hydraulic performance of an ancient Spanish watermill', Renewable Energy, 35(2), pp. 387–396. doi: 10.1016/j.renene.2009.03.033.

Pujol, T. et al. (2015) 'Hydraulic efficiency of horizontal waterwheels: Laboratory data and
    CFD study for upgrading a western Himalayan watermill', Renewable Energy. Elsevier
    Ltd, 83, pp. 576–586. doi: 10.1016/j.renene.2015.04.060.

Rott, N. (1990) 'Note on The History of the Reynolds Number', Annu. Rev. Fluid Mech,
    22(1–1).

Russel, A.W. Vanclay, F. (2007) 'A New Approach to TA From Australia',
    Technikfolgeabschaetzung - Theorie und Praxis, 13(3), pp. 78–79.

Senior, J. (2007) 'Hydrostatic Pressure Converters for the Exploitation of Very Low Head
    Hydropower Potential', School of Civil Engineering and the Environment, p. 204.
    Available at: http://eprints.soton.ac.uk/73702/.

Sigloch, H. (2013) Strömungsmaschinen: Grundlagen und Anwendungen. 5th edn.
    München: Hanser Verlag.

Sommerfeld, A. (1992) Mechanik der deformierbaren Medien. 5th edn. Frankfurt: Harri
    Deutsch Verlag.

Sung, H. and Grilli, S. (2005) 'Numerical modeling of nonlinear surface waves caused by
    surface effect ships: Dynamics and kinematics', Proceedings of the Fifteenth ISOPE
    Conference, 8, pp. 124–131.

UN (2015) The Millennium Development Goals Report 2015.

UN-Water (2015) Water for a sustainable world. doi: 10.1016/S1366-7017(02)00004-1.

UNDP (2015a) Sustainable Development Goals (SDGs).

UNDP (2015b) Sustainable Development Goals Booklet: Introducing the 2030 Agenda for
    Sustainable Development. Paris.

UNIDO (2016) 'World Small Hydropower development report 2016'.

Vashisht, A. K. (2012) 'Current status of the traditional watermills of the Himalayan region
    and the need of technical improvements for increasing their energy efficiency', Applied
    Energy. Elsevier Ltd, 98, pp. 307–315. doi: 10.1016/j.apenergy.2012.03.042.

WWAP (2014) The United Nations World Water Development report 2014. Paris. Available
    at: http://unesdoc.unesco.org/images/0022/002257/225741E.pdf.

Zienkiewicz, O. C. (1971) The Finite Element Method. London: McGraw-Hill.

Zotlöterer, F. (2016) Gravitational vortex machine. Available at: http://www.zotloeterer.com
    (Accessed: 5 November 2016).

Acknowledgements

„Im normalen Leben wird einem oft gar nicht bewußt,
daß der Mensch überhaupt unendlich mehr viel mehr empfängt, als er gibt,
und daß Dankbarkeit das Leben erst reich macht."
Dietrich Bonhoeffer (1906 – 1945)

This thesis began at UNESCO-IHE in Delft on a sunny February day in the Netherlands. Till today, there have been many different weather changes from North Sea chills to persistent drizzle and almost tropical heat waves in Delft. Including Dutch snow storms and rainstorms, between all these temporary uncomfortable spells, there has been mainly sunshine. This thesis, which spans the entire period, would not have been possible without the support and encouragement of many people, especially my family and the UNESCO-IHE community.

First and foremost I would like to express my sincere gratitude to Prof. Arthur Mynett for his enduring patience, support and guidance in the latter stages of this thesis. His presence at UNESCO-IHE, the interaction with him and his inspiration and encouragement has led to the completion of this work. The rudimentary discussions founded a basis for this thesis. Thank you Arthur.

My grateful thanks also to Prof. Nigel Wright for giving me the opportunity to pursue research in the field of small hydropower generation for local solutions. His encouragement and guidance enabled me to keep on track. The EU will certainly miss you Nigel.

My profound thanks to Prof. Christof Wolfmaier and the Esslingen University of Applied Sciences, Germany for enabling me to pursue the work on my thesis. His encouragement and foresight has been a motivation to push forward. Herzlichen Dank Christof.

Thanks to the staff and environment of UNESCO-IHE, who have provided invaluable support in different forms at all times. A special thank you to Jolanda Boots for your ever-efficient administration and a special thank you too to Tonneke Morgenstond for your support in organising a work environment whenever I was at UNESCO-IHE in Delft.

I would like to express my gratitude to the committee members for their interest. I am also grateful to the EU's FP-7 program for the funding of this research under the grant number: 212423.

My final words in the acknowledgements go to my wife Gita with my gratitude to her for her unstinting support at all times. A huge thank you to our children Johannes, Matthias and Katharina: it was a great encouragement from you. Were it not for you, I might have completed this thesis earlier, but without so much cheer along the way. Thanks to you for your support and love at all times.

Pradeep Narrain
UNESCO-IHE, Delft, the Netherlands
October 2017

About the Author

Pradeep Narrain was born in Bangalore, India. He did his schooling at St. Joseph's Boys High School in Bangalore before moving to Germany. He studied Mechanical Engineering at the University of Applied Sciences in Esslingen. After having worked on hydraulic modelling of components for transmission systems in the automobile industry, he started work in the field of CFD.

He has worked on various applied research projects involving external an internal flow modelling using CFD at the Institute of Applied Research "Energetic Systems", Esslingen University of Applied Sciences. The applied research projects included the modelling of radial turbines, optimising exhaust silencers of jet engines for small-scale aircraft turbines. Further projects included CFD simulations of winglets for small aircraft and the design of and nozzles for high pressure water jets for the metal machining industry.

Upon completion of his MSc. in Water Resources and Engineering at the University of Stuttgart in 2009, he moved to UNESCO-IHE at Delft, the Netherlands to begin his PhD on very low head hydropower machines. His PhD focussed on low head hydropower, its importance on a global scale, its contribution to local power generation and the implementation of CFD for flow analysis. While continuing working on his PhD, he returned to Esslingen University in 2012 to work at the Faculty of Automotive Engineering. He is a tutor and laboratory engineer in the Laboratory for Car-Body Engineering. He works on CAD construction methodology of free-form surfaces using the industrial software-packages CATIA V5® and SIEMENS NX®. He is also responsible for investigations of CAD car-body models and design-space in a Virtual Reality environment. He also works on Rapid Prototyping of CAD models and in the field of reverse-engineering, the migration of laser-scan data of clay-models into a virtual/CAD environment.

He is married to Gita Feyl and they have three children: the twins Johannes and Matthias (2007) and their daughter Katharina (2009).

For Product Safety Concerns and Information please contact our EU
representative GPSR@taylorandfrancis.com Taylor & Francis Verlag GmbH,
Kaufingerstraße 24, 80331 München, Germany

Printed and bound by CPI Group (UK) Ltd, Croydon, CR0 4YY
08/05/2025
01864383-0001